Biomedia

Electronic Mediations

Katherine Hayles, Mark Poster, and Samuel Weber, series editors

Biomedia

Eugene Thacker

Electronic Mediations, Volume 11

University of Minnesota Press
Minneapolis · London

Published by the University of Minnesota Press
111 Third Avenue South, Suite 290
Minneapolis, MN 55401-2520
http://www.upress.umn.edu

Library of Congress Cataloging-in-Publication Data

Thacker, Eugene.
    Biomedia / Eugene Thacker.
        p.   cm. — (Electronic mediations ; v. 11)
    Includes bibliographical references and index.
    ISBN 0-8166-4352-0 (alk. paper) — ISBN 0-8166-4353-9 (pbk. : alk. paper)
    1. Molecular biology—Philosophy.   2. Biotechnology—Philosophy.
3. Bioinformatics—Philosophy.   I. Title.   II. Series.
QH506 .T47   2004
303.48'3—dc22   2003017201

Contents

Acknowledgments

Parts of this book were read by or greatly benefited from discussions with a number of friends and colleagues. In particular, thanks are due to Hugh Crawford, Alex Galloway, Diane Gromala, Katherine Hayles, Adrian MacKenzie, Suhail Malik, and Cathy Waldby. Additional thanks to Mark Borodovsky, Bruno Frazier, Nataliya Shmeleva, and Igor Zhulin for their discussions on biology and computing.

Chapter 5 was originally given as a talk at the Modern Language Association convention in 2000; chapter 3 was originally given as a talk at the Society for the Social Studies of Science conference in 2001; and chapter 7 grew out of an informal workshop paper at the Towards Humane Technologies conference at the University of Queensland in 2002. The Nettime list has also provided a forum for experimenting with many concepts in this book.

I would like to express my appreciation to the editors and staff at the University of Minnesota Press, and to the Electronic Mediations series editors. Their experience and support has been instrumental in seeing this book to publication.

Every "acknowledgment" has a sacred space reserved for the impossible. This book could not have been written without Edgar and Moonsoon Thacker, Marie Thacker, and Prema Murthy.

CHAPTER ONE

What Is Biomedia?

Can I Download It?

Cultural anxieties concerning biotechnologies are often played out in the news media, where the latest reports on genetic cloning, stem cell research, gene therapy, and bioterrorism all command a significant amount of reportage on current events. Each of these issues is a discourse of the body, and a discourse that articulates specific kinds of bodies (the body of nuclear transfer cloning, the body of adult stem cells, etc.). The issues they raise are difficult and contentious ones: In which cases should experimentation of human embryos be allowed? In which cases is human cloning warranted? How can engineered biological agents be detected in time to deter a bioterrorist attack? Too often, however, the solutions that come up are haphazard, ad hoc modifications of existing legislation. For instance, the temporary U.S. presidential ban on human embryonic stem cell research in 2000 pertained only to federally funded research, not biotech companies such as Geron.[1] Alternately, when policy changes are made, resolutions usually fall back on more traditional ethical models. An example is the U.S. policy on human embryonic stem cell research, which in essence is a version of "the greatest good for the greatest number of people."[2]

That continued attempts are being made to formulate ethical approaches to such complicated issues is in itself encouraging. Yet, what often goes unquestioned, and uncontested, is whether the foundations on which such decisions are being made is in the process of being transformed. What are the assumptions being made—in policy decisions, in ethical advisory capacities, in research and practice itself—about what constitutes a "body," about how biological "life" is defined, about how emerging biotech fields are affecting our common notions of what it means to have a body, and to be a body? In short, it appears that the debates surrounding issues such as stem cell research—

difficult issues, to be sure—are taking place almost exclusively at the terminal ends of the discourse. The questions most frequently asked—such as "Should we allow research on human embryos?"—do not confront other, more fundamental questions, such as "How are novel techniques developed for mammalian cloning, assisted reproductive technologies, and cellular programming collectively transforming how we think about biological life itself?"

What this means is that, in the case of biotech, political questions are indissociable from questions that are simultaneously philosophical and technical. The philosophical question "What is a body?" is, in biotech, indelibly tied to the technical question "How is a body made?" or, more accurately, "What is a biomolecular body capable of, and how can that be enhanced?" Theories of biological life (the common trope of the "genetic code") are intimately connected to practices of engineering, programming, and design—for instance, in how the genetic code in a test tube is implemented in on-line genomic databases.

In fact, in its current state, we can describe biotech not as an exclusively "biological" field, but as an intersection between bio-science and computer science, an intersection that is replicated specifically in the relationships between genetic "codes" and computer "codes." Areas of specialization in biotech, such as genomics, proteomics, or pharmacogenomics, are each unthinkable without an integrated relationship to computer technologies.[3] Increasingly, a large number of the tools researchers use are not only computer-based, but also Web-based, running from servers housed at universities or research institutes. As industry publications have noted, the traditional "wet lab" of molecular biology is being extended, augmented, and even replaced by the "dry lab" of bioinformatics and computational biology.[4]

This intersection is more than merely technical. It involves the intermingling of two disciplines—computer science and molecular biology—that have traditionally held radically different views on the body.

What Can a Body Do?

To begin with, take two techniques, from two related fields, both of which make use of DNA.

The first technique is protein prediction, and is often referred to as "homology modeling."[5] The field is proteomics, or the study of how DNA sequences produce amino acid sequences, and how those amino acid sequences fold into the complex structures we know as proteins. One way to do this is to make use of a number of computational tools that are part of a growing field known as "bioinformatics." One tool is a Web-based application that lets you input a DNA sequence you are studying, and it then searches several genome databases (also on-line) for candidates for a likely match. The results might contain several DNA sequences or even genes that are close to your sequence. You can then use another tool (also on-line) that will take your DNA sequence, and its potential matches from the genome databases, and then create a profile of the amino acids it is likely to produce. This will be sequence data, and so you will have your original DNA sequence, plus close matches from a genome, and amino acid sequence data (or the raw "protein code"). We have sequences, but no structure yet. We can then take the protein code (the 1D sequence of amino acids) and put them through another tool (yes, on-line) that will, based on existing databases of known proteins, draw up predictions for how different parts of that protein (known as "domains") will fold into different three-dimensional structures (known as "motifs"). A final tool (on-line, but you also have to install a plug-in for viewing 3D molecular models) will collate all those predictions to test for their compatibility as one whole protein structure. If you are lucky, you will get a close match, a 3-D file of the molecular model, and data to the reference model in a database such as the Protein Data Bank. You then have some idea (and a lot of data) about the relationship between your candidate DNA sequence that you started with and the possible proteins or protein domains it may code for (Figure 1).

The second technique is biocomputing, or computing using DNA molecules in a test tube.[6] Also called DNA computing, this technique was developed in the mid-1990s as a proof-of-concept experiment. The concept is that the combinatorial possibilities inherent in DNA (not one, but two sets of binary pairings in parallel, A-T, C-G) could be utilized to solve very specific types of calculations. One famous one is the "traveling-salesman" problem (also more formally called "directed Hamiltonian path" problems): you are a salesman, and you have to go through five cities. You can visit each only once and cannot retrace your steps. What is the most efficient way to visit all five cities? In mathematical terms, the types of calculations are called "NP complete" problems, or "nonlinear polynomial" problems, because they involve a large search field, which gets exponentially larger as the number of variables increases (five cities, each with five possible routes). For silicon-based computers, calculating all of the possibilities of such problems can be computationally taxing. However, for a molecule such as DNA, the well-understood principle of "base pair complementarity" (that A always binds to T, C always binds to G) makes for something like a parallel processing computer, except that it functions not through microelectrical circuits but through enzymatic annealing of single strands of DNA. You can "mark" a segment of any single-stranded DNA for each city (using gene markers or fluorescent dye), make enough copies to cover all the possibilities (using your polymerase chain reaction [PCR] thermal cycler, a DNA Xerox machine), and then mix them into a test tube. The DNA will mix and match all the cities into a lot of linear sequences, and, quite possibly, one of those sequences will represent your most efficient solution to the "traveling-salesman" problem (Figure 2).

The reason for briefly introducing these two techniques is that they are exemplary of what the biomolecular body can do in biotech research. They both use DNA, and they both perform "computational" work in relation to DNA, but there are important differences as well. In a sense, one technique is the inverse of the other: in the first ex-

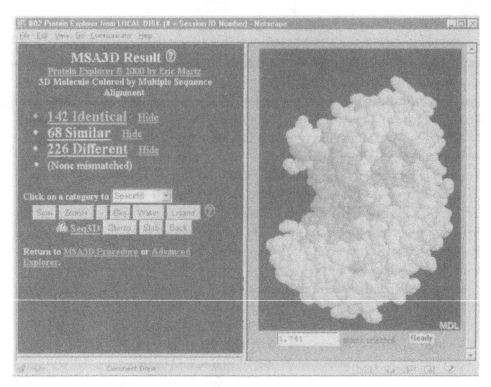

Figure 1. Protein Explorer, an example of a homology modeling application commonly used in bioinformatics research. Reproduced courtesy of Dr. Eric Martz.

ample of bioinformatics, the DNA is fully digital, and the entire process takes place on and between computers; in the second example of biocomputing, the DNA is fully biological, and the entire process of computation takes place in the test tube. Whereas the digital DNA of bioinformatics makes use of computer technology to "model" biology (simulations of "molecular dynamics" in protein folding), the biological DNA of biocomputing is repurposed as a computer in its own right ("base pair complementarity" as two binary sets). The output of bioinformatics is always biological, its point of reference is always the world of the biological cell, the DNA molecule, and various proteins in the body. By contrast, the output of biocomputing is not biological (despite its medium), but rather computational; a "computer" can, theoretically, be made of any material, as long as certain principles (e.g., a storage device, a read program, a write program) are fulfilled.

With these two techniques—protein prediction in bioinformatics, and NP complete calculations in biocomputing—we have a twofold dynamic, in which relationships between biology and technology are significantly reconfigured. On the one hand, there is the premise that biology is computational, that the essence of DNA as a code makes it fully amenable to the digital domain, for archiving, searching, editing, pattern matching, and other computational procedures. It could be said that the success of emerging

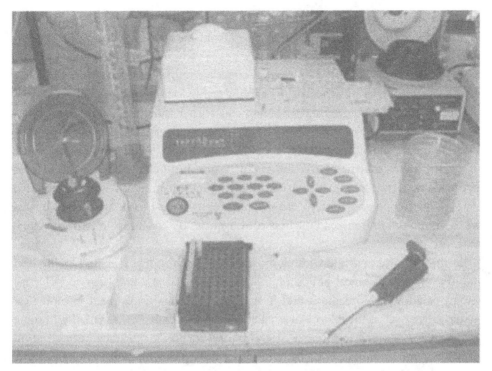

Figure 2. The tools of DNA computing: DNA samples and PCR DNA replication machine. Reproduced courtesy of Dr. Gerald Owenson, University of Warwick, United Kingdom.

fields such as bioinformatics largely depends on the viability of this premise, that we are not simply "representing" DNA, but, in some important way, actually working with DNA as code. On the other hand, there is the premise that computation is biological. Here the emphasis is not so much of computer technology utilized in biological research, but biology utilized in computer science research. The suggestion here is that the characteristics of a "universal computing machine" (in Alan Turing's terms) are such that a variety of material substrates may fulfill a single function.[7]

These two premises—computational biology and biological computing—are informed by a single assumption. That assumption is that there exists some fundamental equivalency between genetic "codes" and computer "codes," or between the biological and digital domains, such that they can be rendered interchangeable in terms of materials and functions. Even though the output of each technique is quite different, they both highlight the centrality of the biological, and its capacity to be instrumentalized into designed contexts. Whether it be the use of biological data in protein prediction (bioinformatics or computational biology), or the use of biological function for non-biological purposes (biocomputing or biological computing), the emphasis is less on "technology" as a tool, and more on the technical reconditioning of the "biological."

It is this assumption, and the twofold logic that extends from it, that characterizes the concept of "biomedia." Put briefly, "biomedia" is an instance in which biological components and processes are technically recontextualized in ways that may be bio-logical or nonbiological. Biomedia are novel configurations of biologies and tech-nologies that take us beyond the familiar tropes of technology-as-tool or the human–machine interface. Likewise, biomedia describes an ambivalence that is not reducible to either technophilia (the rhetoric of enabling technology) or technophobia (the ide-ologies of technological determinism). Biomedia are particular mediations of the body, optimizations of the biological in which "technology" appears to disappear altogether. With biomedia, the biological body is not hybridized with the machine, as in the use of mechanical prosthetics or artificial organs. Nor is it supplanted by the machine, as in the many science-fictional fantasies of "uploading" the mind into the disembodied space of the computer. In fact, we can say that biomedia has no body-anxiety, if by this we mean the will to transcend the base contingencies of "the meat" in favor of virtual spaces.[8]

By contrast, what we find with biomedia is a constant, consistent, and methodical inquiry into this technical-philosophical question of "what a body can do." The ap-parent paradox of biomedia is that it proceeds via a dual investment in biological materiality, as well as the informatic capacity to enhance biological materiality. In some instances, we can refer to this as a "lateral transcendence," or the recontextualization of a "body more than a body." To refer to the examples with which we started: the investment in bioinformatics is not purely digital or computational, but a dual invest-ment in the ways in which the informatic essence of DNA affords new techniques for optimizing DNA through novel software, which in turn makes possible the develop-ment of techniques for enhancing the biological body via new compounds (pharma-cogenomics), new therapies (stem cell therapy), or new diagnostics (patient-specific disease profiling). Biomedia is only an interest in digitization inasmuch as the digital transforms what is understood as biological. In short, the body you get back is not the body with which you began, but you can still touch it. The "goal" of biomedia is not simply the use of computer technology in the service of biology, but rather an empha-sis on the ways in which an intersection between genetic and computer "codes" can facilitate a qualitatively different notion of the biological body—one that is techni-cally enhanced, and yet still fully "biological."

This is a unique configuration of bodies and technologies, made possible as much by technical approaches as by specific technological instruments. In our examples of protein prediction and DNA computing, there are, of course, machines used in the process (e.g., computer and data mining software, PCR cyclers for replicating DNA). But they are not central to the functioning of the novel biomolecular bodies produced through them. The situation is complex enough that it invites a perspective that sees not the machine opposed to the human, and not the artificial opposed to the natural, but a particular instance in which the "bio" is transformatively mediated by the "tech," so that the "bio" reemerges more fully biological. Unlike other biomedical instances of

body–technology meetings—robot-assisted surgery, prosthetic limbs, artificial organs—
the meeting of the body and technology in biomedia is not based on a juxtaposition of
components (human/machine, natural/artificial). Rather, biomedia facilitates and establishes conditionalities, enables operativities, encourages the biological-as-biological.

Although the concept of biomedia presented here is, certainly, open to several interpretations, and more than one application, it is crucial, for a critical understanding of
biotechnology, that biomedia not be confused with "technologization" generally. Biomedia is not the "computerization" of biology. Biomedia is not the "digitization" of
the material world. Such techno-determinist narratives have been a part of the discourse of cyberculture for some time, and, despite the integration of computer technology with bio-technology, biomedia establishes more complex, more ambivalent
relations than those enframed by technological-determinist views. A key component
to the questioning of biotechnology is the attention paid to the ways in which biomedia
consistently recombine the medium of biomolecular systems with the materiality of
digital technology. The biological and the digital domains are no longer rendered
ontologically distinct, but instead are seen to inhere in each other; the biological
"informs" the digital, just as the digital "corporealizes" the biological.

New Media, New Bodies?

These characteristics also point to a significant question: is the juxtaposition of "bio"
and "media" (or "bio" and "tech") not in itself a redundancy? In other words, is the
"body" itself not already a medium? To answer this we can look to two contemporary
formulations of "media" in the context of digital or "new media."

We can begin to address the strange instance of a body that is a medium through a
look at contemporary new media theory. Prior theorists of media such as Walter Benjamin, Martin Heidegger, and Marshall McLuhan have discussed the ways in which
the human subject and the human body are transformed in the interactions with
different technologies. For Benjamin, this primarily takes the form of a novel training
of the senses made possible by the historical-material introduction of media based on
mechanical reproduction, such as film and photography, which has as its primary
effect a detachment and disintegration from an "aura" specific to the object in its time
and place.[9] Heidegger, rather than concentrating on the technical interface between
human and machine, concentrates on interrogating the "essence of technology," which
for him is rooted in an "enframing" that transforms the world into an instrumental
"standing-reserve."[10] For McLuhan, writing some twenty years later, media contain
the capacity to directly interface with the human subject's nervous system, forming an
"extension" of the body, what is at times an almost utopic reconfiguration of the sen-
sorium into a "global village."[11] Contemporary media theorists have carried this notion
further, suggesting that in its more extreme forms, technology fully absorbs the body
(e.g., in posthumanist visions of "uploading"), that technology ambivalently collides

with the body (e.g., Arthur Kroker's study of "panic bodies"), or that technological development configures new boundary arrangements in relation to the body (e.g., Donna Haraway's reappropriation of the militaristic trope of the "cyborg").[12]

In *Remediation*, Jay Bolter and Richard Grusin discuss how "media" and "mediation" have been theorized in ways that have been intimately tied to particular technologies and their broader cultural and experiential impact. They begin with the following: "We offer this definition: a medium is that which remediates."[13] "Remediation" is a concept that describes the ways in which any historically situated media always re-mediates prior media, and thus also re-mediates prior modes of social and cultural modes of communication. For Bolter and Grusin, the concept of remediation involves a complex dynamic between two technological processes: "immediacy" and "hyper-mediacy." The former involves the use of new media to the extent that the media themselves—the "window"—disappear, bringing forth a kind of direct experience where technology is transparent and unnoticed by the subject. By contrast, the latter process—hypermediacy—involves the overcoding, heterogeneity, and saturation of the subject by different media, the empowerment of multiplying media tools at the hands of the subject. As the authors state:

> In addressing our culture's contradictory imperatives for immediacy and hypermediacy, this film demonstrates what we call a double logic of remediation. Our culture wants both to multiply its media and to erase all traces of mediation: ideally, it wants to erase its media in the very act of multiplying them.[14]

In a similar vein, Lev Manovich's study *The Language of New Media* puts forth a series of characteristics of new media that distinguish them from earlier media such as film or television. These are the principles of "numerical representation," "modularity," "automation," "variability," and "transcoding."[15] Among these, it is the concept of transcoding that elaborates the most on the ways in which new media may transform certain visual, haptic, auditory, and corporeal habits specified by earlier media such as film. Technically, transcoding involves all types of file conversion procedures that translate between any two media objects (from still .GIF images to a QuickTime movie, for example). Culturally, this implies a certain universality among hetero-geneous media objects, that a universal code underlies different media, and thus makes possible a horizontally multimediated space. As Manovich states:

> new media in general can be thought of as consisting of two distinct layers—the "cultural layer" and the "computer layer."... Because new media is created on com-puters, distributed via computers, and stored and archived on computers, the logic of a computer can be expected to significantly influence the traditional cultural logic of media; that is, we may expect that the computer layer will affect the cultural layer.[16]

For Bolter and Grusin, a level of transcoding is implied in the very act of remediating; that an earlier medium such as print or film can be remediated in a digital medium such as the Web implies a level of transcoding such that both a print object and a film object can be re-presented in a digital medium such as the Web. Likewise, one of the things that Manovich's characteristics of new media make possible is an unprecedented ability to remediate, permutate, and recombine media elements owing to the technical code-conversion properties of digitization generally. In other words, the concept of "remediation" provides us with one meaning in the "cultural layer" of transcoding. For both Bolter and Grusin, as well as Manovich, the characteristic common to new media is this technical capacity to encode, digitize, and transcode various "things" from the real world (including other media objects).

If this is the case—that is, if it is possible to transcode and remediate various objects from the real world—what effect would this have on the body of the human subject, as an object? Can the body be "transcoded"? Is the body a type of "remediation"?

Bolter and Grusin suggest that the body remediates various cultural and social meanings, while it is also subject to remediation. We may cite examples such as fashion, modern primitivism (piercing, tattooing), body play, cosmetic surgery, transgenderism, bodybuilding, cyborg performance, and other areas of culture as examples of the body both as a medium (a means of communication) and as mediated (the object of communication). As the authors state:

> In its character as a medium, the body both remediates and is remediated. The contemporary, technologically constructed body recalls and rivals earlier cultural versions of the body as a medium. The body as enhanced or distorted by medical and cosmetic technologies remediates the ostensibly less mediated bodies of earlier periods in Western culture.[17]

Bolter and Grusin suggest that any "techniques of the body" situate the body both as a medium and as mediated.[18] Following this, it would seem that cultural attitudes toward the body are the same as those toward media: our culture wants to render the body immediate, while also multiplying our capacity to technically control the body.

However, for Bolter and Grusin—and for many of the theorists they mention—this remediation of the body is an analysis of the ways in which "body" and "technology" are ontologically distinct entities. In such a case, one applies techniques and/or technologies to the body—in cosmetic surgery, for instance—in a way that transforms that body into both a medium (communicating idea[l]s of appearance and aesthetics) and a mediation (a sculpted object of beauty). In other words, in considering the remediated body, something is done to the body in the first place; the body's techniques do not arise from within itself, but rather it gains its remediation externally. One puts on clothing, inserts a piercing, injects ink into the skin, cuts and tucks a section of flesh, attaches prosthetics, utilizes "extensions of the body" such as cellular phones or PDAs in the performance of the everyday.

As a way of diversifying Bolter and Grusin's discussion of the body as remediated, we might ask: what would it mean to approach the body as media in itself? We can begin by considering those instances in which selected properties of the body are geared toward extra-organismic ends. In other words, if we want to inquire into the body as a medium, the first step is to consider the components of the body, along with their range of uses (that is, the relationships between components that constitute the body's range of activity).

That being said, bioscientific fields such as biotechnology have, for some time, focused precisely on this question of the body-as-media: how can selected properties and processes in the biological body be geared toward novel, mostly medical, ends? Specialized research fields in regenerative medicine, genetic diagnostics, and genomics all have this general goal of a nonorganismic technical utilization of the body's biology. Most utilities are for bioscience research and medical application, and others are used in computer science research (DNA computing), drug development (rational drug design), or the materials industries (biomimicry).[19]

As an extension to Bolter and Grusin's discussion of the body (the body both as a medium and as mediated), we can add that a consideration of the body as a medium means that those materialities and processes of technique and technology are not external to or qualitatively different from the body. Whereas the examples raised by Bolter and Grusin—fashion, bodybuilding, cosmetic surgery—rely on both this externalization and this qualitative separation, the body as seen in biotech research generates its technicity from within; its quality of being a medium comes first and foremost from its internal organization and functioning. In earlier techniques such as animal breeding or fermentation, the literal meaning of the term *biotechnology* is indeed this technical utilization of biological processes toward a range of novel ends.[20] The key to the general logic of biotech is that it is not a "technology" in the conventional sense of the term of being a separate instrument; that is, it is not directly concerned with developing external technologies for operating on, controlling, or transforming the "natural/biological" world. But this absence of an identifiable instrument does not mean that instrumentality as such has also disappeared. Although its overall intentions may be congruous with ideologies of technology from industrialism to the so-called computer revolution, biotech is specifically interested in the ways that the material components and biological organization of the body can in effect be reengineered, or redesigned.

To return to Bolter and Grusin's concept of "remediation," we can suggest that a consideration of the body as a remediation also means that it is caught, in its own unique way, between the poles of immediacy and hypermediacy. As an instance of immediacy, the body is situated by the phenomenological concept of "embodiment" or lived experience. However, as an instance of hypermediacy, the body is simultaneously framed by sets of knowledge on the body, including medicine and science. The incommensurability between these—between embodiment and technoscience—is perhaps the zone of the body-as-media. If Bolter and Grusin discuss all media as remediations, we can modulate this statement to suggest that the body is a remediation, a process in which a functioning, biological materiality self-manifests, caught in the midst of the poles of immediacy and hypermediacy, the "body itself" and the body enframed by sets of discourses (social, political, scientific).

Distinguishing Biomedia

This preliminary definition of biomedia—as the technical recontextualization of biological components and processes—is broad enough that it can cover a wide range of practices, from the selective breeding of animals and plants to the everyday use of vitamin supplements or over-the-counter drugs. However, although biomedia may be a labile concept, there are important reasons to argue against such a broad application.

In the broad sense, "biomedia" would simply designate any mediation of the body. This may be technological (as in visual representations of bodies in photography, film, or streaming video), but "mediation" can also be social, cultural, political, or economic. And indeed, we may ask if the "thing" being mediated is not itself already a mediation, as in, for instance, the ways in which a molecular-genetic knowledge of the body affects how we understand our own bodies as part of the processes of embodied subjectivity. The extreme version of this is that dominant scientific paradigms (such as the popular notion of an individual being "written" in his or her genetic code) actually condition subject formation.[21] In such a case, the individual subject, in our particular cultural-historical moment, would arise concurrently with a biologism informed by genetic science. The body of the subject is therefore always already scripted in part by scientific-medical modes of knowledge production. The discomforting suggestion in this version of constructionism is that knowledge of our own bodies is always in some sense at odds with our subjective experience of our bodies.

In this sense, to mediate the body is to reify it twice, first through medicine, second through technical practices, both of which are at odds with our sense of individual subjecthood. The common frustration with such interpretations is that the object of inquiry rapidly fades away beneath the layers of mediation, sometimes taking one to the extreme position of "the body as a sign."[22] But this broad definition of "biomedia" not only infinitely defers any referent we might refer to contingently as "a body," but it also can deny modes of tropic materialization or "corporealization" their real effects in different contexts (medical, social, political). A "cultural constructionism" of the body is not identical to disembodiment or dematerialization; a range of medical, social, technical, and political forces may heterogeneously combine to articulate, say, the "body in medicine" or the "body of immunology," and it is precisely this so-called construction that makes it effective, material, and situated.[23]

Therefore, biomedia does not simply designate the body-as-constructed, for this assumes that "media" are covalent with reification and dematerialization. It also assumes that corporeality or biological materiality is aligned with an extratechnological moment that is consonant with the "body itself." In other words, the broader notion of biomedia as an instance of constructionism formulates an ontological division between the "bio" and the "media," such that the latter has as its main task the mediation of some unmediated "thing." The resultant effect is that the premediated status of the thing is effaced, and placed at a distance from itself. When that premediated thing is taken as the "body itself," what is effaced is a notion of the biological body as natural (the biological as identical to the natural). Most critical formulations, however, do not reduce this relationship to one of originary, premediated body versus reified, mediated body. Instead, the thing mediated is itself seen as a mediation of a qualitatively different order.[24] When, in the process of mediation, the thing mediated is taken as itself a mediation via medical knowledge practices, then what is effaced is not simply nature, but the biological domain as distinct from the technical domain. The biological body—understood as itself a product of medical discourse—is then only understood through its mediation (for example, as in the positioning of medical technologies as diagnostic referents for patient symptoms). The difficulty—one that may never be resolved—is that even the dual mediation of this more critical position still assumes that medical knowledge works upon some thing that it mediates as anatomical, medical, or genetic.

It seems that the assumptions that inform these various critical-constructionist positions ultimately come down to what are philosophical debates over the ontological specificity of biological materiality (organic versus nonorganic matter, animate versus inanimate), or the epistemological verification of materiality itself. In this apparent paradox is perhaps one of the most intimate relationships between the body and language. Responding to the question "Are bodies purely discursive?" Judith Butler states:

> The anatomical is only "given" through its signification, and yet it appears to exceed that signification, to provide the elusive referent in relation to which the variability of signification performs. Always already caught up in the signifying chain by which sexual difference is negotiated, the anatomical is never given outside its terms and yet it is also that which exceeds and compels that signifying chain, that reiteration of difference, an insistent and inexhaustible demand.[25]

Without reiterating such debates, what is useful in considering novel contexts such as those given to us through biotechnology is the "difference" between what phenomenology calls the body and embodiment. Whereas "the body" relates to those social, cultural, scientific, and political codings of the body as an object (public/private, racialized, gendered, anatomical), "embodiment," as Maurice Merleau-Ponty relates, is the interfacing of the subject as an embodied subject with the world (the "flesh of the world").[26] In the context of biology and medicine, this difference is significant, because it relates to the manifold bioethical dimensions in medicine's situating of the patient as a body. This is what Drew Leder refers to as the "dys-functioning" or "absent body": the fact that we often do not notice our body in its autonomic biological functioning until something goes wrong.[27]

Although these phenomenological perspectives are valuable in the embodied differences they point to, we still have a case in which notions of embodiment are rendered ontologically distinct from "nonsubjective" discourses emanating from medicine, and medicine's notion of biological materiality. Furthermore, the condition suggested by the concept of biomedia—and deriving from analyses of emerging fields in biotechnology—is that, in these particular cases, media are indistinguishable from the biological body.

However, the focus here is less on the wide application of the concept of biomedia, and more on a particular, highly significant instance, in which this action of recontextualizing the biological domain is materialized in the contemporary intersections between biotechnologies and computer technologies. We can articulate the concept of biomedia further through the following distinctions:

First, biomedia is not exclusively a concept, nor is it exclusively a technology, a "thing." Biomedia can be thought of as the conditions in which the concept (recontextualizing the biological domain) and the technology for doing so (e.g., bioinformatics tools) are tightly interwoven into a situation, an instance, a "corporealization."[28] Therefore, we can variously speak of biomedia as a single concept (that necessitates a grouping of particular technologies and techniques), and a technology (that requires a certain enabling conceptual organization).

Biomedia is not simply about "the body" and "technology" in an abstract sense, but concerns the biological body situated by a range of "technoscientific" fields. Its specific locale is an interdisciplinary one, in which biological and medical approaches to understanding the body become increasingly indissociable from the engineering and design approaches inherent in computer science, software design, microelectrical and computer engineering, and other related fields. The "body" in biomedia is thus always understood in two ways—as a biological body, a biomolecular body, a species body, a patient body, and as a body that is "compiled" through modes of visualization, modeling, data extraction, and *in silico* simulation.[29] It is this interdisciplinary cross-pollination (biological computing, computational biology) that is characteristic of biomedia.

As a second distinction, we can state that biomedia does not approach "technology" or "media" as fully external to the constitution of the biological domain. For our purposes here, we can take "technology" and "media" as closely related terms, though not exactly identical. They can be taken either as metaphors (a political "technology"; a cultural "mediation") or as designating particular objects (computer technology; visual media). Their main distinction in the context of biomedia will be between technological application and mediation—that is, between their primary modes of functioning. Because instances of biomedia (such as bioinformatics or biocomputing) involve both application and mediation, the two terms will often be understood as being coextensive, or as a technology/media complex.

That being said, biomedia does not approach technology along the lines of the familiar tropes of the tool, the supplement, or the replacement to the human. In the first case, formulated most thoroughly by Heidegger, technology is taken as a tool for the human user, meaning that it is distinct from the body, and comes from the outside to be used by the body, so that it forms a complex of "enframing."[30] This trope is extended in the second trope, where the tool becomes an extension of the body of the user, a supplement that adds new functionalities to the body not inherent in it. This twofold moment of extension and supplementing may also lead to a third trope, where the processes of automation and control lead to technologies that displace or replace the body of the user.[31]

There are a few points to be made here concerning biomedia and technology. One is that biomedia do not so much configure technology along the lines of instrumentality (the common motifs of the tool, supplement, or replacement), although an instrumentalization of the biological body is implicit in the practices of biotechnology. Rather, technology for biomedia is generative, establishing new technical configurations in which the biological can constantly surpass itself. Another point, related to this, is that in biomedia, technique is consonant with technology, in that each field in biotechnology involves a codevelopment of techniques for optimizing the biological, along with an array of technologies that provide for optimizing conditions. This need not be the case in every instance in which a technique or a technological object is involved, but it is the case in the particular fields of biotechnology research under consideration. Technology and technique, however, are never simply applied directly to the biological domain, but instead operate in way in which, as already stated, the biological domain is impelled to function in advance of itself (while still remaining "biological" in its components and processes, though not its context).

Along these lines, biomedia is also not about the relationship between the body and technology as a unilinear dichotomy. Although there are many ways to conceptualize the body–technology relationship, biomedia is based on a particular type of organization. The body–technology relation in biomedia is not based on the substances or composition of the components involved (where carbon-based organisms are opposed to silicon-based computers). Likewise, it is not based solely on functionality (e.g., the artificial organ that functions analogously to the biological body). Finally, it is not based on the technical metaphor of the "interface," in which parts identified according to substance are combined into some hybrid or cyborg-like form (the design engineering principle of "human–machine interface"). Technology is not a "fit" with the body, or vice versa, be this in terms of identifications, functions, or interfaces.

How does biomedia conceive of the body–technology relationship? In part, it does not, and that is perhaps its unique feature. In biomedia, the biological body never stops being biological (we could also say that the biomolecular body never stops being biomolecular); it is precisely for that reason that the biological body is inextricably "technological." This does not, of course, mean that there are no technological objects involved, and no techniques or specific practices. Quite the contrary. But it is how those techniques and technologies are articulated in these biotechnological practices that makes this a unique situation. In biomedia, the "media" employed, and the "technologies" applied, are organized in a way that prioritizes the biological domain as a set of components in interaction with each other via various biomolecular, biochemical, and cellular processes. We could say that technology "enframes" the biological, were it not that this notion implies that technology encloses the biological.[32] The media–technology complex in biomedia creates novel contexts, and establishes novel conditions for biological components and processes. This may involve "technologies" or "media" in the usual sense of the term (from PCR machines to pipettes and petri dishes), and it may involve the techniques that have been standard fare in molecular biology for some time (such as gel electrophoresis, recombinant DNA, or cell culturing).[33] What is different in the case of biomedia is that the use of such technologies, media, and techniques is specifically geared toward enabling the biological domain to technically operate in novel contexts and articulated conditions.

We could say that this is a principles-based approach to the design of the biological domain. A biological principle, such as "base pair complementarity," is recontextualized in the practices of biocomputing, just as the principle of "molecular dynamics" is recontextualized in the process of protein prediction, modeling, and de novo synthesis.[34] This is more than the laboratory replication of a natural environment, for the concerns of biomedia are not so much to simply study the biological domain, but rather to optimize it for a range of uses, both biological and nonbiological.

BmTP: Biomolecular Transport Protocol

Biomedia vary with each particular constellation of biological and technical elements. However, biomedia can also be said to have several characteristics in common. One way of approaching these common characteristics is through the technical principle of "protocols."[35] In computer networking terminology, a "protocol" is a set of rules for establishing how a network operates in a range of contexts. "IP" or "Internet Protocol" is a common type of protocol. Others that may be familiar to Internet users include "FTP" or "File Transfer Protocol" and "HTTP" or "Hyper-Text Transport Protocol." Protocols dictate such things as how a network is operationally configured (do all nodes have direct access to each other?), the relationships between "host" and "client" computers (for instance, in uploading files or downloading e-mail), and the ways in which data is transmitted from computer A at one point on the network to computer B. One of the interesting things about network protocols is that they often employ a flexible set of rules that can adapt to different circumstances. For instance, when viewing a web page, the HTML file with text, images, and any other media are delivered from a host or server computer to a client or the user's computer. However, those files are not delivered all at once. Rather, the IP protocol slices up those files into small chunks, giving each chunk a unique identifier. Each of those little chunks of data is then sent along many different possible routes along the Internet. These chunks of data (not full files, but parts of files) are referred to as "packets." Ideally, they all take the shortest route between two points. However, when Internet traffic is heavy, or when a bottleneck occurs in one of the network connections, the shortest route is often not possible. Therefore, each packet takes one of many alternate routes along the network, sometimes taking detours that may, geographically speaking, span a wide area. This is the reason that, when viewing Web pages, sometimes all of a Web page will load except for one little image, icon, or banner. The Web browser is waiting for the missing packet to literally complete the picture. When all the packets arrive, their unique identifiers tell the browser in which order they are to be reassembled.

If we switch our attention from "new media" to biomedia, we can see a similar protocol at work, but with very different effects and consequences. In short, biomedia can be said to operate according to an informatic protocol, in which the principles for managing data are the means by which biomedia recontextualize the biological domain. This informatic protocol proceeds by a three-step process of "encoding," "recoding," and "decoding."

Encoding

The process of encoding is a boundary-crossing process. When, in speaking of media technology, we say that we are "encoding" (be it scanning digital images, capturing digital video, or coding a text file in HTML), we imply a processual transition from one medium to another, a shift in material substrates. Note that the process of encoding is not identical with demateralization, as it is often taken to be with digital media. The most abstracted, most virtual medium still operates through some material substrate, as all software operates through hardware. Katherine Hayles refers to this as the "materiality of the medium."[36] Therefore, although encoding—or its special case, "digitization"—may be commonly understood as a move from the world of materiality to the world of information, it is important to underscore the complicity of materiality with information.

Encoding, then, is akin to a process of data translation, from one format to another, from one material substrate to another. The key to encoding is, of course, the "difference" between material substrates. If there were no difference between material substrates, then there would be no reason for a translation in the first place. The process of translating data from one format to another—from a photograph to a digital image, from VHS to QuickTime, from .doc to .html—implies a difference that encoding aims to ameliorate. The way in which it does so is through an informatic approach in which "data" is seen to inhere in any organization of matter, any relationship of matter and form. Whereas a more traditional philosophical definition situated matter as formless, and form as realizing matter, the process of encoding adds another dimension, which is that of data.[37] As cybernetics and information theory have suggested, data may not only inhere in matter and form (modeling air, light, or water in computer graphics; 3-D modeling based on simple forms), but data may also be transported across different media. Classical information theory explicitly states—in language and in mathematics—that data may be transmitted from point A to point B, irrespective of the "content" of the data.[38] What makes data different from matter and form? We might say a quantifiable iteration, a persistence in spatialized rhythm, a pattern of relationships—anything consistently peculiar within the matter–form complex that is amenable to quantification (bits, bytes, and data as objects). For the process of encoding, "data" becomes a third term, a trajectory across complexes of matter and form.

What is of interest in the process of encoding is that, on the one hand, it implies a significant difference constituted in part by the materialization of data, that there is literally a "difference that matters" between, say, a photographic print and a digital image on a computer screen. On the other hand, encoding's main reason for being is to overcome this difference (and, arguably, its "matter"), and in the process facilitate a separation between essential data and material substrate. On a philosophical and technical level, encoding raises the question: when a pattern of relationships is transferred from one material substrate to another, does the pattern of relationships remain the same?[39]

One response to this question comes from biomedia. In biotech research, practices of encoding take place daily in the lab: in genome sequencing, in gene expression profiling, in protein analysis and modeling, in digital microscopy, and in cellular diagnostics using silicon chips. At another, more biopolitical level, a different type of encoding takes place in the hospital: the creation, transmission, and modification of medical records, patient-specific data from examinations, computer-based X-ray, MRI, CT, or other scans. But encoding in these contexts does not simply mean translating the body into data, if by this we mean a process of dematerialization. This is so for two reasons: because biomedia's moment of encoding is contingent on a corollary moment of "decoding" (more on this a little later), and because biomedia's informatic approach implies that, in designating patterns of relationships (data), something essential—and functional—is transported from one medium to another.

An example within biotech research is the broad category of "biochips." Research in this field incorporates approaches from mechanical engineering, electrical engineering, computer science, and molecular biology. As we shall see in a later chapter, the hybrid objects produced as part of this research are a paradigmatic example of biomedia. One class of biochip is the "microarray" or the DNA chip. As its name implies, this is literally a tiny silicon chip on which single strands of "sticky" DNA are attached using photolithographic processes. These strands of DNA—thousands are "spotted" on a single chip—stand on their ends like tiny hairs in an aqueous solution. Because they are single stranded, or sticky, they will "naturally" bind with their complementary base when present in solution. Therefore, a sample solution of fragmented DNA of unknown sequence can be washed over the DNA chip. The right base pairs of DNA will automatically attach themselves to their corresponding stands (A and T binding; C and G binding).

Researchers can perform analytic and diagnostic tests on the genetic mechanisms in conditions such as breast cancer or neurodegenerative disorders by using such biochips. If you have a known genetic sequence, such as one from the *p53* tumor supressor gene, you can spot it onto a DNA chip, and give each strand an extra genetic "marker." This marker is really part of a gene that, when bonded with its complementary pair, will cause a fluorescent glow to occur. You can then pass an unknown DNA sample over the DNA chip and see if there are any "hits." If there are, certain spots will glow. This can be organized as a grid, with glowing and nonglowing regions. A specialized computer system can "read" the microarray pattern, according to its pattern of activity, caused by the process of binding single-stranded DNA. Analytic software can perform pattern matching on this microarray output against a microarray database to assess whether this particular DNA test sequence has a close match to any known sequence. The results for this procedure are of two kinds: if there are a significant number of hits on the DNA chip, there is a high likelihood that the test sample is related to your control sample; and, if the test sample returns a match with a microarray database, then there is a high likelihood that the sequence is already known and has been studied. That information can then be utilized to tell you something about your test sample.

This procedure contains many levels of mediation, to be sure. The one we can focus on, however, is the one related to the process of encoding. What is encoded in this use of the DNA chip? In the common sense of the term, nothing is really encoded, and there are no proprietary technologies used in the process: basic tools of molecular biology (generating cDNA libraries; adding gene markers) and basic computer technologies (digital scanning; database queries) are both employed. What we would normally call "digitization" does occur, when the computer scans the microarray pattern as a grid of spots of different colors. Again, this is basic digital imaging, and it would be a stretch to say that the actual DNA is "encoded" in this process.

But what in fact is encoded? It seems that what is encoded is not just an image, but an index, which points to an enframed technical context in which biological processes do or do not occur. Recall that the main principle of the DNA chip is the "natural" process of base pair complementarity. This process occurs, for instance, during protein synthesis, when the tightly woven DNA double strand uncoils and "denatures" itself, so that a single-stranded RNA molecule may be "transcribed."[40] This denaturing and annealing (DNA re-stitching itself) is recontextualized in the DNA chip, not for protein synthesis, but for gene expression analysis. Thus, what we see being encoded is more than just an image of a microarray, but patterns of relationships—base pair binding in DNA that enables DNA diagnostics and that triggers the gene markers (the fluorescent glow) that reiterate the microarray as a grid that can be digitized by a computer, parsed into quantifiable data, and compared against several databases for further analysis.

Recoding

It is at this step of analysis in the computer that the transition from encoding to recoding becomes evident. Not only have certain patterns of relationships been "translated" across material substrates (from the laboratory-prepared test sample, and the

DNA-silicon hybrid of the biochip, to the computer system), but this pattern of relationships is preserved in the new medium; that is, the data that is encoded into the computer is more than just an image. That image is analyzed as a grid, and each grid is a known DNA sequence. Therefore, any "hits" on the grid denote a complementary DNA sequence (you know one side, you know the other). The computer can write those sequences and assemble them into a model sequence of DNA. The model sequence from this original test sequence can then be analyzed using bioinformatics software tools.

This brings us to the second process that defines biomedia, which is "recoding." We are now no longer in the "wet lab" but in a "dry lab," doing "biology *in silico*." And, we have moved from the field of biochips to the field of bioinformatics. If the encoding process carried patterns of relationships across material substrates, then the recoding process will extend the functionality of that encoded or translated data in ways that are specific to its medium.

Like encoding, the term *recoding* also has its common connotations. One major one is that recoding constitutes a form of programming, or better, of reprogramming. For example, in an open-source environment, where a number of people will contribute to a common database of program code (adding code, modulating code, making code more efficient, distributing code), the practice of recoding itself constitutes what is meant by "open source."[41] Recoding raises issues pertaining to the malleability of data, in which data can be positioned along any of the axes that a particular material substrate enables. In the case of open-source initiatives, code can be widely distributed, authored by several programmers, and customized according to very specific activities.

Recoding can be generally thought of as working with data within a context defined by a material substrate (that is, without moving that data to another material substrate). An example within the field of bioinformatics is genetic analysis performed using software tools. If we continue our previous example, we see how the DNA chip enabled an encoding of particular processes (gene expression via base pair binding). Now that these patterns of relationships have been transported to another material substrate (now that they have been digitized), the same genetic test sequence can be worked with in that new medium.

One common technique employed in genetic analysis is "pairwise sequence alignment." Pairwise sequence alignment constitutes one of many techniques used to identify an unknown genetic sequence, by comparing it to databases with known sequences.[42] The overall goal of such techniques is to develop a full "profile" for the test sequence, from characteristics of its sequence to its role in intracellular processes such as protein synthesis or gene expression. With techniques such as pairwise sequence alignment, the test sequence (now in digital form) is arranged in a line, and paired against similar sequences from genome databases. A "scoring matrix" is generated for each sequence comparison, which may be as specific or general as the research dictates. A genetic analysis of this kind actually combines two types of recoding practices: sequence alignment and sequence queries. The latter is often the precondition for the former, in that most sequence alignment analyses depend on being able to use the test sequence to search a genome database. For instance, our *p53* test sequence may be compared to several human genome databases using any software tool that manipulates strings of data (such as FASTA), and that accesses a database and looks for similar sequences (using tools such as BLAST).[43] Analyses can be based on exact identity, sequence similarity (using quantitative methods), or sequence homology (which may combine statistical with evolutionary data on a sequence). The data resulting from a number of pairwise sequence alignments can then be given hierarchical values according to which pairwise alignment returned the best results.

The test DNA sequence can not only be analyzed as a sequence (that is, from the computer's perspective, as strings), but it can also be analyzed according to its potential role in the processes of transcription, translation, and protein synthesis generally. Once a high-scoring alignment is achieved, the test sequence can then be further analyzed by automatically translating it into an amino acid sequence. Another, different set of software tools (such as those at ExPASy) perform such operations as "translation" (from DNA sequence to amino acid code) and "backtranslation" (from a amino acid code into DNA sequence).[44] Because the genetic code is "degenerative" (that is, more than one DNA triplet or "codon" codes for a single amino acid), the user can specify the relations that will constrain the translation process. Once an amino acid sequence (a linear, 1-D protein code) is derived from the test DNA sequence (also 1-D), a range of "structural data" can be generated with these two sequences (DNA and protein). This is, broadly speaking, the interest of the fields known as "structural genomics" and "proteomics": how characteristics of DNA play a part in the structural formation of a 3-D protein. Like the test DNA sequence, the 1-D protein code can now be put through various sequence alignment procedures. However, for the protein code, the data generated will be quite different. Instead of looking for expressed regions in the sequence (as we would with DNA), we are now looking for correspondences between protein sequence and protein structures. Protein structures can be secondary (basic folding classes such as "alpha-helix" or "beta-sheet"), tertiary (combinations of secondary structures together in side chain formations), or quarternary (the entire assemblage of the protein). Using the data generated from the DNA pairwise sequence alignment and the protein sequence alignment, aspects of the 3-D protein structure can be predicted. This "protein prediction" is the same technique with which this chapter began—matching relevant data against known sequence and structural data in databases. With a full profile of a test sample (DNA or protein), a researcher can either identify an unknown biomolecule (gene or protein region), perform further diagnostics studies on already identified biomolecules, or, in the case of medical genetics and pharmacogenomics, isolate candidate "targets" for the development of specific molecular genetics-based therapies.

As is evident from this brief description of bioinformatics techniques, there is a minimal amount of molecular biology work going on, and a great degree of computation. However, despite the thorough use of software tools in this genetic analysis, it would be a mistake to conclude that biology has simply become programming. What is important to note is that the bioinformatics tools, such as those employing sequence alignments (FASTA), database queries (BLAST), or structure prediction (SWISS-PROT), are all developed around biological components and processes. We could be more accurate still and suggest that the aforementioned bioinformatics tools are an example of computer programming creating a context in which certain biological components and processes may function as they do "naturally" in vivo or in vitro. Note that although the material substrate is radically changed (from carbon-based systems to silicon-based ones, from the biological to the digital), what remains the same across this difference are the foregrounded patterns of relationships: transcription (DNA to RNA), translation (DNA to protein), and folding (protein synthesis).

Bioinformatics tools provide a context in which these patterns of relationships are "recoded" as computational integers, algorithms, and sets of rules. It is not enough, in the case of biomedia generally, to say that bioinformatics tools "simulate" the biological body or the cell, for there are no ontological claims being made in these practices of recoding.[45] Rather, the bioinformatics tools, as recoding practices, assume the coexistence of multiple material substrates, and they also assume the capacity for inherent data (patterns of relationships) to be mobilized across those media. In this sense, "recoding" is equivalent to, but not identical with, "wet lab" work with the same patterns of relationships in test tubes, petri dishes, or bacterial plasmids. We would be tempted to state that bioinformatics as a recoding practice claims to be working with the biological domain in itself—but this still assumes an ontological valuation, a displacement of one domain over another. The reason this warrants looking into is that, as we shall see in the next chapter, bioinformatics as a recoding practice bears within itself a tension generated by this character of working with the "essential data" of the biological domain. The very fact that such software tools have been developed for molecular genetics research, and are used on a daily basis, implies that there is much more going on here than a secondary simulation of an originary object. The application of the data generated by bioinformatics tools directly and indirectly "touches" the wet lab (genome sequencing) and the biological body (DNA profiling in medicine, gene therapy clinical trials, genetic drug development). However, when the philosophical question is raised, as it is in the techniques and application of such tools, the implicit assumption is that the digital DNA sequence in the computer points to a corresponding "real" DNA in the lab.

Thus, with the recoding practices of bioinformatics, we see a variable shift in the ontological status of that which is recoded. This is characteristic of biomedia; the ability to continuously modulate the relationships between objects, and the status of individual objects, means that something called "DNA" can simultaneously be split along the lines of instrumentality (digital DNA and bioinformatics as tools for working on real DNA) and fully integrated along the lines of code work (the essential patterns of relationships in both digital and real DNA, dry and wet bodies). In the same way that encoding provides a context for mediation of patterns of relationships across platforms (cross-platform DNA), the practices of recoding extend, modulate, and diversify those patterns in ways that are specific to the medium. The "bio-logic" of DNA is never abandoned—in fact, that is what is preserved as the essential data, the patterns of relationships. But if the bio-logic of DNA is never abandoned, then it is also true that the question "What can a body do?" is answered in ways that are specific to the material substrate (in this case, the digital domain). The "biological" principles of protein synthesis, base pair complementarity, and DNA transcription are still protein synthesis, base pair binding, and transcription. Only now, those patterns of relationships are recoded into medium-specific permutations, enabling "backtranslation," pairwise sequence alignment, and on-line protein folding predictions. From the molecular biological perspective, these are processes that, though based on the same biological principles of DNA, do not occur "naturally" in the living cell. In practices of recoding, we perhaps find one of the central tensions in biomedia: a body that is biological, but a body that is more than biological.

Decoding

As we have stated, the practices of recoding do not exist for biomedia as ends in themselves. The hallmark of biomedia is that the so-called computerization of biology is only legitimized by a third procedure, that of "decoding," in which practices of recoding reach ahead to their prior material substrate, and in doing so effect a qualitative shift in the components, processes, and relations with which we began. The form of the pathway here is less a linear path or a loop, and more of a spiral. In many specific instances— such as "rational drug design," gene therapy based on sequence insertion, and programmable stem cell differentiation—an originary wet biological sample is encoded, then recoded, and finally decoded, so that that sample may be transformed (using the principles established by genetic engineering) or so that new contexts may be developed (using the principles from more recent fields, such as "regenerative medicine").[46]

If encoding elicits associations with digitization (but is really a transport of data across media), and if recoding elicits associations with programming (but is really an extension of data specific to a medium), then decoding would appear to elicit associations that have to do with cryptography (encryption, decryption, code breaking, code translating). At the basis of cryptography and practices of decoding one finds the productive process of making sense, that is, making "information" from "noise."[47] Decoding does not make sense from nothing, but rather works with prior material in a combinatory, even stochastic, method, in order to generate sense from the combinations of parts (literary uses of anagrams are the "poetic" equivalent of this technical approach). However, decoding, as a making sense from prior material (in-formation from noise), has specific connotations when we consider it in relation to biomedia.

What would it mean, given our discussion thus far, to "decode" the body? In one sense, the body is decoded and decrypted on a daily basis, as is evidenced by the role of body language, urbanism, consumerism, fashion, and a myriad of other bodily modes. In biotechnology's molecular view however, the decoded body is something significantly unique. Aside from the popular notions of a "genetic code" and efforts to "crack the code," both of which self-consciously reference cybernetics, the decoded body in this case is also a rematerialized, rebodied body. In many ways, decoding can be regarded as the converse of encoding. Both employ a common technique, which is the isolation of certain types of essential data (patterns of relationships), and the mobilization of that data across media (material substrates). In the case of encoding, we saw how this was specified as a move from the "wet lab" to the "dry lab." In the case of decoding, this relationship is reversed: the essential data from a dry lab context (the particular medium of the computer) is mobilized to a wet lab context (the particular medium of an engineered bacterial plasmid or cells in culture). Notice that this is not simply a move "back" to the wet lab from the dry lab. This is for two reasons. The first is that decoding does not necessarily select the same essential data that encoding does. What is considered "essential" about the data may be in principle the same (e.g., preserving protein synthesis of a particular biomolecule), but in the process of recoding, what constitutes that particular principle may be different (e.g., specific gene expression patterns that cannot be "naturally" observed). This leads to the second reason we do not simply go "back" to the wet lab, which is that recoding has redistributed the types of medium-specific relationships that may occur, such that a design approach to molecular biology can afford novel products (e.g., proteins) from familiar processes (e.g., gene expression).

Furthermore, we not only do not go "back" to a starting point, but decoding is also not simply a direct "materializing" of what was once immaterial. If we accept that encoding is not a dematerializing (even in cases of "digitization") but a rematerializing, then the same would hold for decoding practices as well. There is no originary object in biomedia; but this does not mean that materializing procedures do not play an important role as mediating functions. In fact, if we were to condense our description of biomedia, we could describe it as biotechnical contexts in which the biomolecular body is materialized as a mediation: materialization is the medium.

An example is, within biotech research, the broad area defined as "pharmaco-genomics." Also known as "rational drug design," this field is identical with research associated with pharmaceutical corporations, whose primary aim is to mine the data from genomes and proteomes to develop novel "drug targets" for possible testing, clinical trials, and, if lucky, actual product development (along with patents and official approval by governmental regulatory organizations). As with any pharmaceuticals field, rational drug design involves coming up with novel compounds that in themselves do not exist in nature, though their components may. And, as with a great deal of biotech research, rational drug design depends on computer technology to enable the efficient, "intelligent," and reliable discovery of potential drug targets. Rational drug design can proceed along a number of lines, but its general premise is that, in the long run, synthetically produced drugs in the lab should be replaced by novel genes and proteins in the body that can produce such compounds on their own.[48]

This involves a combination of traditional drug design (analysis of compounds and synthesis of "responses" to those compounds) as well as genome and proteome analysis. The protein prediction techniques alluded to earlier are here utilized in a combinatorial fashion, to test the viability of different compounds. For instance, in the testing of a novel drug therapy that involves a protein inhibitor that will bind to a specific antigen, a number of biochemical and biophysical considerations are computed by a software program, to check for potential nonviable atomic force repulsions. It will then further analyze the compound by performing "molecular docking" tests to assure the correct three-dimensional "fit" between antibody and antigen. If a likely candidate is developed, it can then be cross-checked against a protein database for similar structures, which will in turn lead to a search against a genome database for potential genes or gene sequences that play a role in the production of a particular protein. These candidate genes can then be analyzed, and, potentially, modified in a way that may produce the desired novel protein. If a biological connection can be established between a candidate gene and a designed compound, then the main goal of rational drug design becomes one of creating the right types of contexts in which that connection may be functionally (that is, biologically) established. In the recent past, gene therapies have been used in this way, inserted into the human patient's genome, to either supplement or inhibit certain biochemical processes.[49] However, because a great number of processes that occur in the cell are not unilinear or even remotely centralized processes, gene therapy has had a great deal of trouble—and tragedy—in clinical trials.[50] Another end application aside from gene therapy has been the use of rational drug design in "custom-tailored drugs." When combined with same-day genetic profiling of patients, rational drug design has been hoping to use practices of decoding to design and administer drug therapies specific to the genome of each individual patient.

In these techniques at least two decoding processes are occurring. The first is between the computer-modeled novel protein and its potential extension in the wet lab. Here, patterns of relationships that exist in the digital domain (which are themselves encoded from molecular dynamics studies in the biological domain) are "extended" from the molecular docking software to the web lab bioreactor in which the protein may be synthesized. A second process is nearly the reverse of the first, and that is the extension of particular candidate sequences from a genome database into the wet lab, in order to establish a biological connection between novel compound and candidate gene sequence. Again, it is not simply the sequence itself that is extended from the digital to the biological domain; in this process of decoding, it is the capacity for a candidate sequence to function as a gene (that is, transcription, translation, gene expression) that enables its decoding across media, across platforms. Of course, not all patterns of relationships are viable across media; only those that are viable across media will fulfill the aim of biomedia, which is to enable a cross-platform compatibility for selected patterns of relationships. When a novel compound is synthesized in the wet lab, made possible by analyses using computer and bioinformatics tools, which themselves use data that are encoded from prior wet lab analyses, then the process of decoding signals a final turn in the spiral of biomedia.

Rational drug design proceeds by step-by-step analysis, testing, and modulation of drug candidates. It is, by and large, dependent on the data, and the general premise of the genome—that DNA sequences of variable length called "genes" play the central role in the production of protein or polypeptide compounds. The identification of a "source code" here opens the biomolecular body to the principles of "design." Design in this context implies the configuration of an open-ended set of approaches and principles whose aim is to lead to the most optimal integration between form and function, sequence and structure, decoding and the capacity to be upgraded. "Design" in this context is also not to be immediately taken as a term with automatic moral designations when applied to living beings; that is, design applied to living systems should not be immediately interpreted here as "dehumanizing." Although we will consider the relationships between bioethics and design in a later chapter, for the time being we should first consider design in the context of biomedia's informatic protocol of encoding, recoding, and decoding. It may be said that, in biomedia's technical recontextualization of the biological domain, design often comes to play the role of the "tonality" of the recontextualizing process. Increasingly, living systems are no longer viewed in biotech research through the mechanist lens of engineering, but are approached from the perspective of design, and design as both a set of principles and an approach, an attitude, rather than an application of sets of discrete techniques.

In our discussion of this "informatic protocol" that constitutes biomedia, we can note a twofold tendency, which, arguably, is a product of this intersection between bio-science and computer science in biotechnology research. The first tendency is that of establishing a cross-platform compatibility, by selecting certain types of essential data or patterns of relationships in a given context (e.g., DNA translation into protein). The implication here is that the medium does not matter, because that same pattern of relationships can be implemented in a variety of media (e.g., protein synthesis in a petri dish, in a bioinformatic application, or in drug design). In this sense, "data" is not specific to the digital domain, but is something that inheres in any organization of matter. However, this also leads to the second tendency, which is that the medium does in fact matter. A given pattern of relationships will take on significantly different characteristics given a different material substrate or technical context (e.g., "natural" protein synthesis replicated in a petri dish versus multiple "impossible" pairwise sequence alignments between DNA and protein codes). Biomedia—in our specific cases of biotech research—therefore asks two mutually implicated questions: First, what is a body? The response here is that the body is biomedia, in that its inherent informatic quality enables a cross-platform compatibility. Second, what can a body do?[51] The response here is that the affordances of a given body are in part contextualized by the particular material substrate through which they are articulated.

Multi-, Hyper-, Bio-

Given these characteristics of biomedia, we can now ask about the kinds of bodies generated through such practices of encoding, recoding, and decoding. Such bodies are not "bodies" in the sense of anatomical or even anthropomorphic bodies (a traditional biological standpoint, with an emphasis on the composition of parts in relation to a whole), and such bodies are also not "bodies" in any mechanistic sense (be it a "clockwork machine" or a cyborgic, human–machine "interface"). In fact, biomedia does not produce "things" so much as it generates the conditions in which the tension between patterns of relationships and material substrates takes on several general characteristics.

One of these characteristics is that the notion of enabling a cross-platform compatibility runs parallel with the enabling of "passages" between genetic "codes" and computer "codes." As we shall see in chapter 2, this passage between genetic and computer codes is the foundation that informs much of molecular genetics, though it has been transformed significantly between the postwar era and the new millennium. The establishing of a passage between genetic and computer codes is based on the assumption that some essence or essential data pervades biomolecular bodies such as DNA or proteins, which enables them to be compositionally and functionally transported from one medium to another, or which, at the very least, enables them to be metaphorically enframed as units that operate according to an informatic logic. The genetic "code" is not only a trope but also a database, and the passage between computer and genetic codes is not only a back-and-forth mobility, but also one in which code comes to account for the body (e.g., genetic profiling), just as the body is biotechnically enabled through code practices (e.g., genetic drug therapy).

This total passage across platforms is not, it should be reiterated, a pure investment in the supposed dematerialized domain of the digital domain. Such claims have characterized research associated with "posthumanism," or "extropianism," in which the liberatory promises of new technologies (such as artificial intelligence [AI], robotics, nanotech, and smart drugs) lead the way to a kind of utopian perfection of "the human" through life extension, intelligence augmentation, and the science-fictional concept of "uploading" (or replicating one's mind into software systems).[52] Biotechnologies generally are included in the more long-term, utopian visions of the posthuman, but in a very specific manner. Whereas the general movement in utopian-posthuman thinking is away from the contingencies of the biological domain (or "the meat"), biotechnology is not always a transcendentalist practice. This is evident if we consider biotech through the lens of biomedia's protocol (encoding-recoding-decoding). As we have suggested, biomedia is not so much about the digitization of the biological as it is about enabling certain types of data to be mobilized across different media. While biomedia illustrates the abstraction necessary for this movement across platforms, it also demonstrates the affordances that a given medium advances to that abstracted data. Thus, DNA is not only DNA whether it is in a test tube or database, but it also takes on different dimensions, depending on the context of its mediation (e.g., DNA as protein synthesis, or DNA as a biocomputer). This dual emphasis on abstraction-materialization is central to biomedia. In the discourse of the posthuman, it means that biomedia is not purely a transcendentalist practice, and that, instead of moving away from the contingencies of the biological domain, biomedia moves toward them. This is to say that there is no body-anxiety in biomedia's informatic protocol; the practices of encoding, recoding, and decoding are geared both to move across platforms and to always "return" to the biological domain in a different, technically optimized form. Biomedia returns to the biological in a spiral, in which the biological is not effaced (the dream of immortality), but in which the biological is optimized, impelled to realize, to rematerialize, a biology beyond itself. This is made possible by the movement across platforms, which is itself a condition of the establishment of passages between genetic and computer codes.

There are several consequences to such instances. One is that the body is accounted for through data, and that access to the body, access to "knowledge of the body," is provided for by the context of biomedia. Not only does this affect individual instances such as genetic disease profiling in the clinic, but it also affects research broadly, for instance, in the ways in which a universalized "human genome" as well as various population genomes come to account for the biomolecular body of the universal patient of medical genetics.

The accountability for the body through information also means that the context of biomedia (its informatic framework) facilitates certain approaches, techniques, and research questions at the expense of other approaches and modes of critical inquiry. Medicine—at least in the envisioned medical practice imagined by the biotechnology research community and industry—will move a long ways from the prior models of Galenic bedside medicine, early-twentieth-century holistic medicine, or homeopathy, and, in its emphasis on the mediations between genetic and computer codes will emphasize a biomolecular body that is open to biological optimization.

In this sense, the viewpoint of bioinformatics gives us a radically different notion of biological normativity and "health." Biology can, in some ways, be read as a refrain of the question "What is life?"[53] Different researchers, at different periods, working within different disciplinary contexts, answer this question in their specific ways. But one common trend is that biological "life" has some essential relationship to the notion of information, be it purely metaphoric or imminently technical. Increasingly, the field of molecular genetics research is demanding that the biologist also be a computer scientist.

This is borne out by the increase in job positions and industry productivity in the field of bioinformatics; a number of subfields are almost unimaginable without computer science (structural genetics, high-throughput genomics, proteomics, comparative genomics). We might say that, from the perspective of biomedia, biological life is largely becoming a science of informatics, but an informatics whose existence, practices, and worldview are predicated on the enabling of mobility across platforms (that is, across material substrates).

At this point, it will be useful to consider some of the basic assumptions in biomedia, as it is manifested in biotech research fields such as genomics, bioinformatics, proteomics, and medical genetics.

There is, in biomedia, a general devaluation of material substrates (the materiality of the medium) as being constitutive of patterns of relationships (or essential data). Although biomedia do take into account the role of material substrates, they exist, as we have seen, in the backdrop, as support for "what the body can do" (or, more accurately, what patterns of relationships can do). Again, biomedia generally and biotechnology specifically are not dematerializing technologies, at least in the posthumanist or Extropian senses of the term. Biomedia is constantly working toward the body, always coming around via a spiral, and enframing this movement as a return to the body. The difference within this spiral movement, the difference that makes it a spiral and not a loop, is the tension between abstract essential data (patterns of relationships) and the media (material substrate) in which that data inheres. However, as we have seen in the examples of bioinformatics, biocomputing, microarrays, protein prediction, and rational drug design, the materiality of the medium literally matters, a difference that makes a difference. Biomedia is not only predicated on the ability to separate patterns of relationships from material substrates, but, in never completely doing away with the material orders, it often relegates the constitutive qualities of material substrates to the role of delivery, of a vehicle for data, of transparent mediation.

To return to our definition: biomedia is neither a technological instrument nor an essence of technology, but a phenomenon in which a technical recontextualization of biological components and processes enables the biomolecular body to demonstrate itself, in applications that may be biological or nonbiological. In its informatic protocol of encoding, recoding, and decoding, biomedia bears within itself a fundamental tension. On the one hand, there is the ability to isolate and abstract certain types of essential data, or patterns of relationships, which are independent of and mobile across varying media, or material substrates. On the other hand, something is implicitly added through these varying media, such that the essential data never remains completely untouched, but is itself becoming infused and in-formed by the integration with the medium. This is further complicated by the fact that, with biomedia, the aim or application is not to move beyond the material substrate, but to constantly appear to return to it, in a self-fulfilling, technical optimization of the biological, such that the biological will continue to remain biological (and not "technological").

As contemporary biotechnology research pursues the integration of bio-science and computer science, it might be worth continuing to think through the philosophical-technical implications that arise from the biomedia that are generated (such as protein prediction software and DNA computers).

"New ways, new ways, I dream of wires"

The gap alluded to at the beginning of this chapter—between emerging biotechnologies and the bioethical discourses that attempt to address their impact—seems to be a clear sign that alternative ways of viewing the body, biological life, and the relationship between human subjects and technologies are needed. This book is an attempt to address this gap, through a focus on the philosophical-technical aspects of emerging fields in biotechnology. Although this does not imply that the analyses are without any relation to cultural and political aspects of biotech, what has been foregrounded here are the ways in which our varying notions of "the human," biological "life," and the boundary between the body and technology are being transformed within biotech research. It should be stressed that this study is not explicitly a critique of globalization, capital, and biotech; other individuals and groups are approaching this issue in sophisticated and politically engaged ways.[54] Nor is it concerned with the broader dissemination of biotechnology as a "cultural icon," for instance, in popular culture, mainstream media, or in instances of the everyday.[55] It is, however, an inquiry into the difficult questions that biotech inadvertently puts forth in its techniques, technologies, and concepts, and in this sense the main methodological focus here is simultaneously philosophical and technical.

In this concern, the individual essays intentionally take up "emerging" research fields as case studies—bioinformatics, biocomputing, MEMS (microelectromechanical systems) research, nanomedicine, and systems biology. These are fields that may be interdisciplinary, involved in research that is as much theoretical as empirical, and that, in some cases, have no definite area of projected application. They are also situated as "future sciences" or "fiction sciences" through their rhetoric (e.g., articles in magazines such as the *MIT Technology Review* and *Wired*), and as components of a biotech industry (start-up companies, new job markets, new degree programs).[56] The primary reason for considering such fields is to ask how these questions concerning the body, biological life, and the body–technology boundary are addressed and resolved in explicitly technical ways.

The focus on these fiction sciences is also not without a strong link to science fiction (or, SF). Certainly, the implication is not that science is "fictional," if by this we mean nonexistent, ineffectual, or a pure construction of cultural, social, or economic forces. However, as emerging fields, such areas as biocomputing and nanomedicine do participate in a particular type of discourse, a talking and doing, in which projected applications, development in knowledge, and relationship to other disciplines and industries are all employed in a speculative fashion. Any field, in attempting to distinguish itself as a field, will by necessity look around itself, to state what it is not, or which fields it combines in an interdisciplinary manner. Involved in this gesture is a speculation of what the field could be, in the same way that it is standard for research articles to speculate, in either their opening or closing remarks, the future directions and significance of their empirical data. For this reason, each chapter begins with an anecdote from selected works of contemporary science fiction. The goal of this is not to derealize technoscientific practice, but in fact to render it all the more "real."

The informatic protocols of encoding are the focus of chapters 3 (on biochips) and 4 (on biocomputing). Both biochips, such as DNA microarrays, and biocomputers, such as DNA computers, involve the construction of literal hybrids between living and nonliving components and processes. Both are also examples of encoding practices, which highlight patterns of relationships and transport them into novel contexts. However, their overall aims are significantly different, roughly being divided between medical and computational application.

Likewise, the practices of recoding are foregrounded in chapter 2, which deals with bioinformatics, as well as chapter 6, which examines alternative approaches in biotech research broadly known as "systems biology." The central question in these examinations of encoding is whether systems-based approaches (focusing on "biopathways" and studies of "systemic perturbations") might transform the very assumptions in biomedia's informatic protocol—the notion of equivalency between genetic and computer codes.

Finally, the practices of decoding are the focus of chapter 5, on nanotechnology and nanomedicine. Whereas a number of these biotechnology fields reconfigure the relationships between bodies and technologies, nanotech has for some time implied that, at the atomic level of matter, the distinction is itself irrelevant. Specialized medical applications in nanotech research are materializing such claims, in the design of in vivo biosensors and nano-scale devices for medical diagnostics. As a practice of decoding, nanomedicine demonstrates how patterns of relationships and material substrates are one and the same.

This study closes with a final chapter on the relations between notions of "design" and bioethical issues raised by biotechnologies. However, bioethics in this sense is taken to be a practice fundamental to research in progress and research as process, rather than post hoc responses to finished research. This chapter examines the ways in which the question of design suddenly takes on contentious moral and ethical resonances when design is imbricated in the biological domain, and especially the human body. The questions we close with have to do with the possibility for a technically sophisticated "actant bioethics" that places as much emphasis on the "affect" of design as it does on the design of embodied practice, or "enaction."

Finally, it should be stated that two general, open-ended, "impossible" questions run throughout this study:

First, what is "biomolecular affect"? We are used to thinking of affect and phenom-enological experience generally in anthropomorphic terms. Is there a phenomenology of molecular biology? Are there zones of affect specific to the molecular domain and irreducible to "molar" aggregations or anthropomorphisms? What would such an analy-sis say concerning our common notions of embodied subjectivity?

Second, is the body a network? Although work in immunology and certainly neu-rology has answered this in the affirmative, the work from molecular genetics is more ambivalent. Is the biomolecular body a distributed relation? If so, what does this say concerning more traditional, centralized notions of consciousness, "mind," and the normative anatomical-medical enframing of the body?

Whereas the Extropian and posthumanist dreams of "uploading" continue a move away from the body and embodiment, there is, perhaps, another type of posthuman, even inhuman, body within these two questions.

CHAPTER TWO

Bioinformatics

BLAST, BioPerl, and the Language of the Body

Morphologies

Diaspora, a science-fiction novel by Greg Egan, opens with the birthing of a "citizen" of the thirtieth century.[1] As may be guessed, neither this citizen nor the birthing process is what we would expect. A "citizen" is a sentient life-form that exists entirely through computer systems and networks, belonging to a "polis." The polis—a robust, future version of a server computer—may exist in some remote location, such as in deep space or buried far beneath an Earth desert, using wireless and satellite links to transfer citizens to any location on the network. Each polis, as its name indicates, is a type of governing organization, with its own set of rules and social interactions among citizens and between polises.

That being said, the birth of a citizen in the future world of *Diaspora* is as far from biological reproduction as the citizens are from biological bodies. Because the citizens of a polis exist as sentient software, or as code with consciousness, their processes of "birth" are more akin to designed iterations of informatic patterns, where a desired pattern "emerges" from a set of simple rules:

> The Konishi mind seed was divided into a billion fields: short segments, six bits long, each containing a simple instruction code . . . The conceptory's accumulated knowledge of its craft took the form of a collection of annotated maps of the Konishi mind seed . . . But there was one single map which the citizens of Konishi had used to gauge the conceptory's progress over the centuries; its showed the billion fields as lines of latitude, and the sixty-four possible instruction codes as meridians. Any individual seed could be thought of as a path which zig-zagged down the map from top to bottom, singling out an instruction code for every field along the way.[2]

Interestingly enough, the tropes used by the polis still reflect the biological—and cultural—aspects of biological reproduction, where a code-generating computer "womb" runs genetic algorithms on a "seed," which would develop into a "psychoblast" until it reached the stage of self-awareness, at which time it was assigned citizenship in the polis. In the informatic future world of *Diaspora*, a protocitizen is therefore not born, but rather generated through recursive, positive feedback loops of genetic algorithms operating on "flesher" DNA patterns:

> In the orphan psychoblast, the half-formed navigator wired to the controls of the input channels began issuing a stream of requests for information. The first few thousand requests yielded nothing but a monotonous stream of error codes; they were incorrectly formed, or referred to non-existent sources of data. But every psychoblast was innately biased toward finding the polis library . . . and the navigator kept trying until it hit on a valid address, and data flooded through the channels.[3]

The AI-based approaches in "machine learning" are here commingled with the "innate" DNA patterns that emerge through iterative processes, an embryological development of the genetic body through the "ports," "addresses," and networks. *Diaspora* opens with this scene in particular because it portrays a biologically "impossible" birth, an "orphanogenesis," or generation of a citizen from no parents. The orphan, named Yatima, becomes one of the central characters in the novel, which follows several story lines that depict posthuman life in varied forms.[4] What is particularly interesting about this scene, however, is the condensation of the biological and the informatic, of genetic "codes" and computer "codes." Two aspects of the process Egan describes are derived from current research in computer science: genetic algorithms and DNA databases. The latter is a central aspect of the nascent field of "bioinformatics," which has not only transformed the way in which molecular genetics and biology research is framed, but which has also played an indispensable role in the mapping of the human genome. Bioinformatics may be simply described as the application of computer science to molecular biology research. The development of biological databases (many of them on-line), gene-sequencing computers, computer languages (such as XML-based standards), and a wide array of software tools (from database queries to gene and protein prediction) are all examples of the ways in which bioinformatics is literally transforming the face of the traditional molecular biology laboratory. Some, more optimistic, reports have suggested that the "wet" lab of traditional biology is being replaced by the "dry" lab of "point-and-click biology."[5]

The scenes depicting the generation of sentient software in *Diaspora* are, of course, extrapolations of science and technology—in this case from contemporary research in computer science, AI, complexity, and biotechnology. However, what *Diaspora* imagines is a situation in which an implosion of genetic and computer codes becomes the occasion for the emergence of nonbiological "life"—yet a life that is based on the patterns of information specific to DNA, itself a particular type of pattern of biomolecules in a cell nucleus. This successive layering of genetic and computer codes serves as a science-fictional example of how the organization of genetic and computer codes leads to "generative code," or the technical articulation of a life-form that is both genetic and computational. Although the "citizens" of *Diaspora* are both disembodied and nonbiological, it is worth noting that what has been preserved from the fleshers is not substance, but rather pattern—DNA. Thus, when we speak of the intersections of genetic and computer codes, we are not necessarily talking about a division between body and machine, organic and nonorganic, natural and artificial, for this belies the complex ways in which "the body" has been enframed by genetics and biotechnology. In the example from *Diaspora,* the conceptory actually does nothing; rather, it is the infrastructure of the mind seed—particular patterns of genetic code indissociable from computer code—that feeds back upon itself, thereby generating the stochastic process of designed software akin to genetic algorithms.

In short, *Diaspora* presents us with an instance in which a technical recontextualization (in this case, of the abstract "pattern" of DNA) provides the space in which the logic of DNA acts upon itself in computational ways. Again, this iterative layering of genetic and computer codes is characteristic of contemporary biotech fields such as bioinformatics. Although current uses of bioinformatics are mostly pragmatic (that is, technology-as-tool), the issues that bioinformatics raises are much more fundamental; that is, although bioinformatics is predominantly a technical field, such technical fields also raise a series of largely philosophical questions concerning biological "life" and the relation between biology and technology (or, to use another pair, between the "natural" and the artificial). The questions raised bear upon our common assumptions concerning the differences between representations (or simulations) and the "thing itself" in molecular biology. They also have to do with the question of how this intersection of genetic and computer codes may be transforming the way in which bioinformatic bodies of all types are approached—sequences, molecules, databases, software, and cells. Finally, they have to do with the possibility that the broad influence of bioinformatics in research, application, and, ultimately, medical practice may significantly alter notions of what constitutes a "body" in terms that are scientific, philosophical, social, and technical.

Bioinformatics in a Nutshell

The use of computers in biology and life-science research is nothing new, and, indeed, the informatic approach to biological life may be seen to extend back to the development of statistics, demographics, and the clinic during the eighteenth century.[6] The first computer databases used for biological research are closely aligned with the first attempts to analyze and sequence proteins. In the mid-1950s, molecular biochemist Fred Sanger published a report on the protein sequence for bovine insulin.[7] This was accompanied a decade later by the first published paper on a nucleotide sequence, a type of RNA from yeast.[8] Other like projects followed, filling in research on both nucleotide and protein sequences of selected organisms, from the Drosophila fruit fly to various bacteria.[9] From a bioinformatics perspective, this research can be said to have been instrumental in identifying a unique type of practice in biology: the "databasing" of living organisms at the molecular level. Note that these examples do not involve computers, but rather take an informatic approach, in which, using wet-lab procedures, the organism is transformed into a sequence database.

The reports of the first nucleic acid sequencing research were immediately followed by efforts to gather, organize, and annotate the data being generated. Because most of the data was related to proteins (structural data derived from X-ray crystallography), the first bioinformatic computer databases were built in the 1970s (Protein Data Bank) and 1980s (SWISS-PROT database).[10] With the increased availability of computer workstations at universities, selected tools for working with this data appeared in the 1980s, and, with the addition of the personal computer and the Web, those tools have multiplied since. Bioinformatics was, however, given its greatest boost from the original Human Genome Project (HGP), a joint Department of Energy and National Institutes of Health (NIH)–sponsored endeavor based in the United States, and initiated in 1989–90.[11] The HGP went beyond the single-study sequencing of a bacteria, or the derivation of structural data of a single protein. It redefined the field, not only positing a universality to the genome, but also implying that a knowledge of the genome could only occur at this particular moment, when computing technologies reached a level at which large amounts of data could be dynamically archived and analyzed. The HGP put to itself the problem of both regarding the organism as itself a database and porting that database from one medium (the living cell) to another (the computer database) (Figure 3).

"Bioinformatics" as an identifiable discipline can be said to grow out of these trajectories. Bioinformatics, as its name indicates, is an intersection of molecular biology and computer science in bioscience research. As one researcher states, bioinformatics is specifically "the mathematical, statistical and computing methods that aim to solve biological problems using DNA and amino acid sequences and related information."[12] However, bioinformatics is also a business. In 2000, Silico Research Limited reported that commercial bioinformatics software was worth close to $60 million, with many biotech and pharmaceutical companies outsourcing a quarter of their R&D to bioinformatics start-ups.[13] The investment firm Oscar Gruss estimates that bioinformatics' market—as a subset of the software industry—could be as much as $2 billion by 2005.[14] In addition, bioinformatics bridges different industries through its applications, prompting a number of information technology (IT)–bioinformatics collaborations: IBM's "Blue Gene" supercomputer for protein folding, Compaq's "Red Storm" supercomputer, developed in collaboration with Celera Genomics, and the involvement of Hitachi, Motorola, and Sun in developing IT tools for bioinformatics.[15]

As both a set of practices and an industry, bioinformatics has three main sectors. The first is software development, with companies such as Spotfire, MDL, and Silicon Genetics providing stand-alone applications (often sold as suites), and others, such as eLab and eBioinformatics, offering Web-based tools. Second, the software development side is also connected to the development and management of databases and hard-

Figure 3. Production sequencing facility at the U.S. Department of Energy's Joint Genome Institute in Walnut Creek, California. Reproduced courtesy of U.S. Department of Energy Human Genome Program, http://www.ornl.gov/hgmis.

ware systems. Companies such as Lion Bioscience, InforMax, and Perkin-Elmer specialize in either database management or the development of hardware systems for genome sequencing. Third, software and the databases they interact with will run into bottlenecks unless a data standard can soon be established. A number of companies are dedicated to "ontology" standards, or developing cross-platform file format, annotation, and markup languages for all of the data being generated in bioinformatics research. Consortiums of biotech companies, such as the I3C (Sun, IBM, Affymetrix, Millennium Pharmaceuticals, and others), are working toward such standardizations of bioinformatics as a practice.

What do researchers do with bioinformatics tools? Molecular biologists can perform five basic types of research using bioinformatics tools: digitization (encoding biological samples), identification (of an unknown sample), analysis (generating data on a sample), prediction (using data to discover novel molecules), and visualization (modeling and graphical display of data).

Among the most common techniques is sequence alignment, also one of the earliest in bioinformatics. Also referred to as "pairwise sequence alignment," this simply involves comparing one sequence (DNA or protein code) against a database for possible matches or near matches (called "homologs"). Doing this manually is not only exhausting to think about, but also impossible, for it involves pattern matching along several "reading frames" (depending on where one starts the comparison along the sequence). Bioinformatics tools perform such pattern-matching analyses through algorithms that accept an input sequence, and then access one or more databases to search for matches.[16]

Other common techniques include sequence assembly (matching fragments of sequence), sequence annotation (notes on what the sequence means), data storage (in dynamically updated, on-line databases), sequence and structure prediction (using data on identified molecules), microarray analysis (using biochips to analyze test samples), and whole genome sequencing (such as performed by human genome projects).[17] Bioinformatics tools can be as specific or as general as a development team wishes. They can search databases for open reading frames, exon splice sites, repeat sequences, single nucleotide polymorphisms (SNPs), protein motifs and domains, expressed sequence tags (ESTs), and candidates for PCR primer design.[18] When combined with molecular modeling tools, bioinformatics can also be used to study protein structure, as well as to aid in the design of compounds for genetic drugs or therapies.

Command-Control-Restart

If one wanted to get an idea of these dual biological and technical strands of bioinformatics (computer science plus bioscience), it would be possible to do so through a look at the various pop-science books authored by researchers over the years. We could begin, for example, with Erwin Schrödinger's *What Is Life?*, one of the earliest formulations of genetic material in terms of "information" and "codes" (though at the time of its publishing the structure of DNA and the coding schema had not been elucidated).[19] We could also include Francis Crick's *Life Itself*, as well as his numerous articles on "the genetic code" throughout the 1950s and 1960s.[20] These writings were largely responsible for popularizing genetics's "central dogma" inside and outside of molecular biology circles. Likewise, biologists George and Muriel Beadle's *The Language of Life* summarized the findings of molecular biology not only in informatic terms, but also in linguistic terms, helping to promulgate the idea that DNA was in some sense a "language."[21] Popular books by the French team of François Jacob and Jacques Monod extended the informatic tropes of DNA by adding to the discourse their famous research on genetic regulatory mechanisms.[22] Jacob and Monod took the notion of DNA-as-information further, suggesting that gene regulation formed a "genetic program" in itself that was not unlike the mainframe computers then being developed by the U.S. military and businesses such as IBM. In fact, it seemed that, throughout the late 1950s and 1960s, a number of molecular biology and genetics researchers published books or articles for a nonspecialist public, each of which was a version of Schrödinger's question "What is life?" and each of which was a version of the answer: biological life is DNA, and DNA is information.

As Lily Kay effectively demonstrates, the trope of the genetic code has had a long life, but one with fits and starts, divergent paths, and discontinuous paradigms.[23] Kay's overarching point is that the notion of the "genetic code"—and indeed molecular genetics itself—emerges through a cross-pollination with the discourses of cybernetics, information theory, and early computer science. Kay provides roughly three discontinuous, overlapping periods in the history of the genetic code—a first phase marked by the trope of "specificity" during the early part of the twentieth century (where proteins were thought to contain the genetic material), a second "formalistic" phase marked by the appropriation of "information" and "code" from other fields (Watson and Crick's research fits in here, especially Crick's formulation of "the coding problem"), and a third "biochemical" phase during the 1950s and 1960s, in which the informatic trope is extended, such that DNA is not only a code but a fully fledged "language" (genetics becomes cryptography, as in Marshall Nirenberg and Heinrich Matthai's work on "cracking the code" of life) (Figure 4).

Although Kay's historical analysis concludes with this third phase (her book stops in the 1960s, just prior to genetic engineering), it is not difficult for us to see at least two phases following upon it. One would have to be a "biotechnical phase," in which the ability to study and analyze genetic processes leads to new ways of manipulating, regulating, and controlling those processes—a "control principle" in genetic engineering. The development of recombinant DNA techniques in the early 1970s is largely seen to spawn the era of genetic engineering, and the first international concerns over the ethical use of emerging biotechnologies.[24] Herbert Boyer and Stanley Cohen's recombinant DNA research had demonstrated that DNA not only could be studied, but could be rendered as a technology as well. The synthesis of insulin—and the patenting of the techniques for doing so by Genentech—provides an important proof of concept for biotechnology's control principle in this period.[25]

The most recent phase—a "bioinformatic phase"—has to do less with genetic engineering's control principle and more with the integration of computer science and biotechnology. The race to map the human genome proved in the end to be about bioinformatics more than anything else—the key players in the race were not scientists but supercomputers, databases, and programming languages. Even in its inception in the late 1980s, the Department of Energy's Human Genome Project signaled a shift from a control principle to a "storage principle," while never forgoing the ability to control genetic matter that characterized genetic engineering.[26] This bioinformatic phase is increasingly suggesting that biotech and genetics research is nonexistent without some level of computer technology. Emergent fields are each accompanied by a novel technology: gene expression (biochips), genomics (automated sequencing computers), proteomics (supercomputers), structural genomics (data mining software).

Note these are not sequential but concurrent phases; though the genetic engineering tools of the biotechnical phase develop earlier than bioinformatics tools do, the latter are unthinkable without the agenda of the former. The control principle of genetic engineering has taken new forms with the availability of on-line genomic databases, just as the storage principle in bioinformatics has enabled the proliferation of novel software tools.

FIRST BASE	SECOND BASE: U	SECOND BASE: C	SECOND BASE: A	SECOND BASE: G	THIRD BASE
U	UUU Phe	UCU Ser	UAU Tyr	UGU Cys	U
U	UUC Phe	UCC Ser	UAC Tyr	UGC Cys	C
U	UUA Leu	UCA Ser	UAA Stop	UGA Stop	A
U	UUG Leu	UCG Ser	UAG Stop	UGG Trp	G
C	CUU Leu	CCU Pro	CAU His	CGU Arg	U
C	CUC Leu	CCC Pro	CAC His	CGC Arg	C
C	CUA Leu	CCA Pro	CAA Gln	CGA Arg	A
C	CUG Leu	CCG Pro	CAG Gln	CGG Arg	G
A	AUU Ile	ACU Thr	AAU Asn	AGU Ser	U
A	AUC Ile	ACC Thr	AAC Asn	AGC Ser	C
A	AUA Ile	ACA Thr	AAA Lys	AGA Arg	A
A	AUG Met	ACG Thr	AAG Lys	AGG Arg	G
G	GUU Val	GCU Ala	GAU Asp	GGU Gly	U
G	GUC Val	GCC Ala	GAC Asp	GGC Gly	C
G	GUA Val	GCA Ala	GAA Glu	GGA Gly	A
G	GUG Val	GCG Ala	GAG Glu	GGG Gly	G

Figure 4. The genetic code. Triplet codons each code for a particular amino acid. Reproduced courtesy of Dr. Robert Huskey.

Broadly speaking, we can summarize this historical overview by describing a twofold movement: that between metaphorization and autonomization. In the early and mid-century phases articulated by Kay, we have the gradual and discontinuous process by which the concept of "information" in cybernetics and information theory is appropriated by molecular biology as a metaphor for describing the genetic material (what Kay calls a "metaphor of a metaphor").[27] DNA acts in a manner analogous to information in technical systems, being both the carrier and the message from one generation to the next.

However, in the phases that follow upon this—the biotechnical and bioinformatic phases—information is not taken as a metaphor for DNA, but is seen to inhere in DNA itself as a technical principle. With the rise of novel techniques for controlling and storing DNA in computers, the metaphoric stature of the informatic model collapses onto DNA itself. The development of genome databases and bioinformatics software tools seems to point to the fact that information is no longer a metaphor for DNA, a way of talking about DNA, but that, technically speaking, DNA *is* information. This autonomization of DNA with respect to information means that the genome itself can be regarded as a biological computer—something demonstrated in non-biological uses of DNA in biocomputing, for instance.[28]

This broad transition from a metaphorization to an autonomization of DNA as information does not, of course, imply that biotechnology currently operates without metaphors; the recent hype surrounding the mapping of the human genome attests to the contrary—metaphors of the "book of life" and the "software of the self" were abundant, in both specialist and nonspecialist media. What it does imply is that biotechnology research no longer looks to other fields for its metaphoric inspiration. Rather, in assuming the informatic tropes of DNA, biotech research increasingly approaches biological problems with the general principles of informatics. "Information" no longer comes from the outside (disciplinarily speaking) to describe a biological entity such as DNA. Rather, information is seen as constitutive of the very development of our understanding of "life" at the molecular level—not the external appropriation of a metaphor, but the epistemological internalization and the technical autonomization of information as constitutive of DNA. This cannot be overstated. It is this assumption that enables biotech research to envision a genome that can be variously encoded (sampled into an on-line database), recoded (data mining for novel genes), and decoded (synthesis of novel drug compounds): the body is no longer a medium transmitting the information of DNA (the model from cybernetics and information theory); the body is itself "biomedia." Anytime we encounter this technical recontextualization of biological processes, we have an instance of biomedia. No steam machines, no cyborgic fusions of metal and flesh, no Turing machines, machine intelligence, or black boxes— only the technical design of the conditions in which "natural" biological processes can occur in novel contexts.

Again, we would be wrong in thinking that a sufficiently advanced technology makes metaphor unnecessary; the current "bioinformatic phase" of the genetic code is not about the ability of technology to illuminate the vagueness of metaphor. But the over-all status of the informatic metaphor with regard to biotech's understanding of "life" has significantly changed, and this is, in part, owing to the ways in which new computer technologies have been integrated into biotech research. Although researchers of an earlier time—such as Crick—never proposed the application of the principles of information theory, cybernetics, or electrical engineering to molecular biology research, their interpretation of "information" from those fields did significantly transform molecular biology—and the kinds of questions it could ask. In short, we can say that the metaphor of the "genetic code" has served a largely descriptive role within molecular biology, but it has gone on to become internalized as an epistemological foundation for biology and genetics. What we have been witnessing in the past twenty years or so (with the rise of a biotech "industry") is a further level at which the metaphor of information is materialized through sets of practices, techniques, and technologies. The genome database is but one example of this (description, internalization, materialization). Indeed, nowadays it seems nearly impossible to think of DNA, or the genome, or biological life itself outside of those informatic terms established during the postwar era. When such approaches are offered, they are often dismissed as either too theoretical, nonscientific, or even mystical. The theories of symbiosis, epigenetics, complexity, and autopoiesis have all, at one time or another, been the target of such critiques.[29]

Given such histories, what are we to make of contemporary "What is life?" books such as Pierre Baldi's *The Shattered Self,* a book written by a bioinformatician, which suggests that the new biotechnologies are challenging our very notions of selfhood?[30] On the one hand, it appears, uncannily, that researchers in bioinformatics are unconsciously beginning to put poststructuralist theories of the fragmented subject into practice (into code...). But Baldi's argument gives evidence not from a philosophical-cultural perspective, but from a biological-technical one: as Baldi suggests, the more we learn about "life" at the molecular and genetic levels (at the informatic levels), the more we find that our common notions of an autonomous, unified, atomistic "self" are incommensurate with the view from biotech and bioinformatics:

> In fact, our notions of self, life and death, intelligence, and sexuality are very primitive and on the verge of being profoundly altered on the scale of human history... This shattering is brought about by scientific progress in biology, computer science, and resulting technologies such as biotechnology and bioinformatics.[31]

Although Baldi attributes the deconstruction of the modern subject to scientific progress (which largely implies progress in computer science), we may question this combination of conservatism and progressivism. This biomolecular view of life presents us with a "body" that is, in a troubling way, irreducible to the various anthropomorphisms that are attributed to the genome or DNA. A given process, such as the metabolism of sugar, may involve a network of thousands of biomolecules in networks and sub-networks. This "biomolecular body" is both human and inhuman, at once constituting ourselves as organisms and forming a bio-logic that has very little to do with representation, reductionism, or agency/causality. From this perspective, the challenge facing biotech today is not getting enough computational power, but asking the question of whether our notions—philosophical and ethical—of "the body" and "the human" are in need of qualitative reformulations.

"We're in the building where they make us grow"

Amid this description of bioinformatics practices, one question might beg asking: aren't we simply dealing with data, data that has very little to do with the "body itself"? In many ways we are dealing with data, and it is precisely the relation of that data to particularized bodies that bioinformatics materializes through its practices. We can begin to address this question of "just data" by asking another question: when we hear talk about "biological data" in relation to bioinformatics and related fields, what exactly is meant by this?

On one level, biological data is indeed nothing but computer code. The majority of biological databases store not biological nucleic acids, amino acids, enzymes, or entire cells, but rather strings of data. In computer programming terminology, "strings" are simply any linear sequence of characters, which may be numbers, letters, or combinations thereof. Programs that perform various string manipulations can carry out a wide array of operations based on combinatorial principles applied to the same string. This may be calculations on numbers, or permutations of letters in text. In fact, the most familiar type of string manipulation takes place when we edit text. In writing e-mail, for instance, the text we type into the computer must be encoded into a format that can be transmitted across a network. That means that the statement "In writing e-mail, for instance, the text we type into the computer must be encoded into a format that can be transmitted across a network" must be translated into a lower-level computer language of numbers, each group of numbers representing letters in the sequence of the sentence. At an even more fundamental level, those numbers must themselves be represented by a binary code of zeros and ones, which are themselves translated as pulses of light along a fiber-optic cable (in the case of e-mail) or within the microcircuitry of a computer processor (in the case of word processing).

In text manipulations such as writing e-mail or word processing, a common encoding standard is known as ASCII, or the American Standard Code for Information Interchange. ASCII was established by the American Standards Association in the 1960s, along with the development of the Internet and the business mainframe computer, as a means of standardizing the translation of English-language characters on computers. ASCII is an "8-bit" code, in that a group of eight ones and zeros represents a certain number (such as "112"), which itself represents a letter (such as lowercase "p"). Therefore, in the sentence "In writing e-mail, for instance, the text we type into the computer must be encoded into a format that can be transmitted across a network," each character—letters, punctuation, and spaces—is coded for by a number designated by the ASCII standard.[32]

How does this relate to molecular biotechnology and the biomolecular body? As already noted, most biological databases, such as those housing the human genome, are really just files that contain long strings of letters: As, Ts, Cs, and Gs, in the case of a nucleotide database. When news reports talk about the "race to map the human genome," what they are actually referring to is the endeavor to convert the order of the "strings" of DNA in the chromosomes in the cell to a database of digital strings in a computer. Although the structural properties of DNA are understood to play an important part in the processes of transcription and translation, for a number of years the main area of focus in genetics and biotech has, of course, been on DNA or "genes." In this focus, of primary concern is how the specific order of the string of DNA sequence plays a part in the production of certain proteins, or in the regulation of other genes. Because

Table 1. Correlation between DNA, ASCII, and binary code

DNA	A	T	C	G
ASCII	65	84	67	71
Binary	01000001	01010100	01000011	01000111

sequence is the center of attention, this also means that, for analytic purposes, the densely coiled, three-dimensional, "wet" DNA in the cell must be converted into a linear string of data. Since nucleotide sequences have traditionally been represented by the letter of their bases (adenine, cytosine, guanine, thymine), ASCII makes for a suitable encoding scheme for long strings of letters. To make this relationship clearer, see Table 1.

In the same way that English-language characters are encoded by ASCII, the representational schemes of molecular biology are here also encoded as ASCII numbers, which are themselves binary digits. At the level of binary digits, the level of "machine language," the genetic code is therefore a string of ones and zeros like any other type of data. Similarly, a database file containing a genetic sequence is read from ASCII numbers and presented as long linear strings of four letters (and can even be opened in a word-processing application). Therefore, when, in genetics textbooks, we see DNA diagrammed as a string of beads marked "A-T" or "C-G," what we are looking at is both a representation of a biomolecule and a schematic of a string of data in a computer database.

Is that all that biological data is? If we take this approach—that is, that biological data is a quantitative abstraction of a "real" thing—then we are indeed left with the conclusion that biological data, and bioinformatics, is nothing more than an abstraction of the real by the digital.[33] This may be the case from a purely technical perspective, but we should also consider the kinds of philosophical questions that this technical configuration elicits; that is, if we leave, for a moment, the epistemological debate of real versus digital, the thing itself versus representation, and consider not "objects" but rather relationships, we can see that biological data is more than binary code. Take, for example, a comment from the Bioinformatics.org Web site:

> It is a mathematically interesting property of most large biological molecules that they are polymers; ordered chains of simpler molecular modules called monomers. Think of them as beads or building blocks which, despite having different colors and shapes, all have the same thickness and the same way of connecting to one another... Many monomer molecules can be joined together to form a single, far larger, macromolecule which has exquisitely specific informational content and/or chemical properties. According to this scheme, the monomers in a given macromolecule of DNA or protein can be treated computationally as letters of an alphabet, put together in pre-programmed arrangements to carry messages or do work in a cell.[34]

This is to suggest that the notion of biological data is not about the ways in which a real (biological) object may be abstracted and represented (digitally), but instead about the ways in which certain "patterns of relationships" can be identified across different material substrates, across different platforms. We still have the code conversion from the biological to the digital, but rather than the abstraction and representation of an object, what we have is the cross-platform conservation of specified patterns of relationships. These patterns of relationships may, from the perspective of molecular biology, be elements of genetic sequence (such as base pair binding in DNA), molecular processes (such as translation of RNA into amino acid chains), or even structural behaviors (such as secondary structure folding in proteins). The material substrate may change (from a cell to a computer database), and the distinction between the wet lab and the dry lab may remain (wet DNA, dry DNA), but what is important to note is how "biological data" is more than just abstraction or representation. This is because the biological data in computer databases is not merely for archival purposes, but is there as data to be worked on, data that, it is hoped, will reveal important patterns that may have something to say about the genetic mechanisms of disease.

Although a simulation of DNA transcription and translation can be constructed on a number of software platforms (including 3-D and computer graphics–based output), it is important that the main tools utilized for bioinformatics begin as database applications and Unix-based text-processing tools. What these particular types of computer technologies provide is not a more perfect representational semblance, but a medium-specific context, in which the logic of DNA can be conserved, and played out in various experimental conditions. What enables the practices of gene or protein prediction to occur at all is a complex integration of computer and genetic codes. What must be conserved is the pattern of relationships that is identified in wet DNA (even though the materiality of the medium is transferred). A bioinformatics researcher performing a multiple sequence alignment on an unknown DNA sample is interacting not just with a computer, but with a "bio-logic" that has been conserved in the transition from the wet lab to the dry lab, from DNA in the cell to DNA in the database.

Biological data can be described more accurately: the consistency of a "bio-logic" across material substrates or varying media. This involves the use of computer technologies that both conserve the bio-logic of DNA (e.g., base pair complementarity, codon-amino acid relationships, restriction enzyme splice sites) and do so by developing a technical context in which that bio-logic can be recontextualized in novel ways (e.g., gene predictions, homology modeling). In this way, bioinformatics is constituted by a challenge, one that is as much technical as it is theoretical: it must regulate the "difference" between genetic and computer codes.

On a general level, biotech's regulation of the relationship between genetic codes and computer codes, carbon and silicon, would seem to be mediated by the notion of "information" as a distinct element that is capable of accommodating differences across media, the establishment of a third principle that is able to form correlations between heterogeneous phenomena. This "translatability" between media—in our case, between genetic codes and computer codes—must then also work against certain transformations that might occur in translation. Thus the condition of translatability (from genetic to computer code) is not only that linkages of equivalency are formed between heterogeneous phenomena, but also that other kinds of relationships are prevented from forming, in the setting up of conditions whereby a specific kind of translation can take place.[35] We can further articulate this bio-logic in several ways.

A bioinformatics approach to the body not only involves the preservation of informational purity across media, but, more important, the disabling of transformation across different media. This is analogous to the approach of "noise reduction," in which the fidelity of genetic data is maintained through transmission, file conversion, and between various hardware platforms (from an on-line server to a locally run sequencer). In this informatic context, the term *noise* carries its classical connotation of any distortion of an information signal at the origin point A, as it is transmitted to destination point B.[36] For bioinformatics approaches, this means not only that genetic data must be capable of being "translated" between molecules and bits, but also that genetic data must remain self-identical in a variety of research-based contexts. A researcher encoding a sample of DNA from a bacterial cell to a genome database must be sure that, in the process, no changes will inadvertently occur to what that researcher deems essential (e.g., raw DNA sequence). Thus, for bioinformatics approaches, the technical challenge is to effect a "translation without transformation," to preserve the integrity of genetic data, irrespective of the media through which that information moves. Above all, the noise reduction in the process of translation is concerned with a denial of the transformative capacities of different media and informational contexts themselves. For bioinformatics, the medium is not the message; rather, the message— a genome, a DNA sample, a gene—exists in a privileged site of mobile abstraction, which must be protected from the specificities of different media platforms.

An additional preventive measure in bioinformatics involves the predilection in research for "stable media," or the concerns, from the perspective of systems operators, for creating robust, consistent, and predictable computer systems for working on genetic data.[37] In Bruno Latour's discussions of science practice and "nonhuman actants," the seemingly passive, inert, inactive objects of laboratories, technologies, and artifacts are instead considered as active participants in the production of knowledge and facilitation of creative thinking in the sciences.[38] Such active objects, or actants, variously put up resistances, reroute actions, or have transformative effects on both the humans and nonhumans involved. Although Latour's arguments apply to technoscientific research broadly, what is specific about this emphasis on stable media in bioinformatics is the way in which it differentiates itself from the traditions of the "wet" molecular biology lab. Linked to the computer hardware and software industries, bioinformatics must constantly manage cycles of technical development (and obsolescence), as well as efforts toward standardization of file formats, operating system platforms, and various other technical features.[39] In part because biotech approaches digital technologies-as-tools, and in part because such technologies are taken as transparent (transparent

so that the biological body can be apparent), biotech demands a level of technical sta-
bility in its engagement with digital technologies—computers, databases, software,
laboratory tools, and so forth. It demands that the actants of biotech research function
as inert, tool-like objects, and not as transformers of the meanings of material practice. The DNA chip is not supposed to display its hybrid, even monstrous, molecular
stitchings of engineered DNA molecules and silicon base; as a diagnostic tool, it is
constructed so as to operate as a neutral template, against which a given cell sample
can be "decoded." Thus, from a technical standpoint, the qualities of instability, unpredictability, accidents—the qualities of "unstable media"—must be regulated if
there is to be a smooth translation between the genetic and the informatic.

Bio-Logic

We can take a closer look at how the bio-logic of techniques and technologies in bioinformatics operates, by considering the use of an on-line software tool called "BLAST,"
and the technique of pairwise sequence alignment referred to earlier.

BLAST is one of the most commonly used bioinformatics tools. It stands for "Basic
Local Alignment Search Tool," and was developed in the 1990s at the National Center
for Biotechnology Information (NCBI).[40] Like many bioinformatics tools, BLAST performs analyses on sequence data, or, in computational terms, on strings. In particular,
BLAST, as its full name indicates, is a set of algorithms for conducting analyses on sequence alignments. The sequences can be nucleotide or amino acid sequences, and the
degree of specificity of the search and analysis can be honed by BLAST's search
parameters. The BLAST algorithm takes an input sequence and then compares that
sequence to a database of known sequences (for instance, GenBank, species-specific
databases, EST databases, restriction enzyme databases, protein databases). Depending on its search parameters, BLAST will then return output data of the most likely
matches. A researcher working with an unknown sequence can use BLAST in order to
find out if the sequence has already been studied (BLAST includes references to research
articles and journals when there is a match), or, if there is not a perfect match, what
"homologs" or close relatives a given sequence might have (BLAST also includes near
matches based on a statistical parameter set by the user). Either kind of output will tell
a researcher something about the probable biochemical characteristics and even function of the test sequence. In some cases, such searches may lead to the discovery of
novel sequences, genes, or gene-protein relationships.

Currently, the NCBI holds a number of different sequence databases, all of which
can be accessed using BLAST, and using different versions of BLAST. For instance,
"blastn" is the default version, searching only the nucleotide database of GenBank.[41]
Other versions perform the same basic alignment functions, but with different relationships between the data: "blastp" is used for amino acid and protein searches, "blastx"
will first translate the nucleotide sequence into amino acid sequence, and then search
blastp, "tblastn" will compare a protein sequence against a nucleotide database after

translating the nucleotides into amino acids, and "tblastx" will compare nucleotide sequence to a nucleotide database after translating both into amino acid code. In addition, when BLAST first appeared, it functioned as a stand-alone Unix-based application, requiring researchers to not only have a working knowledge of the Unix environment, but also to learn BLAST-specific commands.[42] With the introduction of the Web into the scientific research community in the early 1990s, however, BLAST was ported to a Web-ready interface front end and a database-intensive back end. Other bioinformatics tools Web sites, such as the University of California at San Diego's Biology Workbench, will also offer portals to BLAST searches, oftentimes with their own front end.[43]

BLAST has become somewhat of a standard in bioinformatics because it generates a large amount of data from a relatively straightforward task. Though sequence alignments can sometimes be computationally intensive, the basic principle is simple: comparison of strings. First, a "search domain" is defined for BLAST, which may be the selection of a particular database. This constrains BLAST's domain of activity, so that it does not waste time searching for matches in irrelevant data sets. Then an input sequence is "threaded" through the search domain. The BLAST algorithm searches for closest matches based on a scoring principle. When the search is complete, the highest scoring hits are kept and ranked. Because the hits are known sequences from the database, all of their relevant characterization data can be easily accessed. Finally, for each hit, the relevant bibliographical data is also retrieved. A closer look at the BLAST algorithm shows how it works with biological data (Figure 5).

In considering a popular bioinformatics tool such as BLAST, it is important to keep in mind the "control principle" and the "storage principle" of bioinformatics. BLAST unites two crucial aspects of bioinformatics: the ability to flexibly archive and store biological data in a computer database, and the development of diversified tools for accessing and interacting with that database. Without a database of biological data—or rather, a database of "bio-logics"—bioinformatics tools are nothing more than pure software. Conversely, without applications to act on the biological data in the database, the database is nothing more than a static archive. Bioinformatics may be regarded as this ongoing attempt to seamlessly integrate the control and storage principles across media, whether in genetically engineered biomolecules or in on-line biological databases.

As the BLAST algorithm in Figure 5 illustrates, the question of integrating the control and storage principles requires more than pure computer science; it requires an analogous integration of molecular biology with programming, the development of a functional bio-logic. Although databases are not, of course, exclusive to bioinformatics, the BLAST algorithm is tailored to the bio-logic of nucleotide and protein sequence.[44] The bio-logic of base pair complementarity among sequential combinations of four nucleotides (A-T; C-G) is but one dimension to the BLAST algorithm. Other, more complex aspects, such as the identification of repeat sequences, promoter and repressor regions, and transcription sites, are also implicated in BLAST's biological

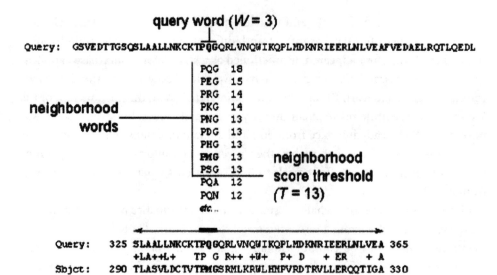

Figure 5. The BLAST search algorithm. Reproduced courtesy of the National Council on Biotechnology Information (NCBI).

algorithm. What BLAST's algorithm conserves is this pattern of relationships, this bio-logic that is seen to inhere in both the chromosomes and the database.

But BLAST not only translates the biomolecular body by conserving this bio-logic across material substrates, but, in this move from one medium to another, also extends the control principle to enable novel formulations of biomolecules not applicable to the former, "wet" medium of the cell's chromosomes. One way in which BLAST does this is through the technique of the "search query." The "query" function is perhaps most familiar to the kinds of searches carried out on the Web using one of many "search engines," each of which employs different algorithms for gathering, selecting, and ordering data from the Web.[45] BLAST does not search the Web, but rather performs user-specified queries on biological databases, bringing together the control and storage principles of bioinformatics. At the NCBI Web site, the BLAST interface contains multiple input options (text fields for pasting sequence or buttons for loading a local sequence file), which make use of special scripts to deliver the input data to the NCBI server, where the query is carried out. These scripts, known as "CGI" or "Common Gateway Interface" scripts, are among the most commonly used scripts on the Web for input data on Web pages. CGI scripts run on top of HTML Web pages, and form a liaison for transmitting specific input data between a server computer and a client's computer.[46] A BLAST query will take the input sequence data and send it, along with instructions for the search, to the server. The BLAST module on the NCBI server will then accept the data and run its alignments as specified in the CGI script.

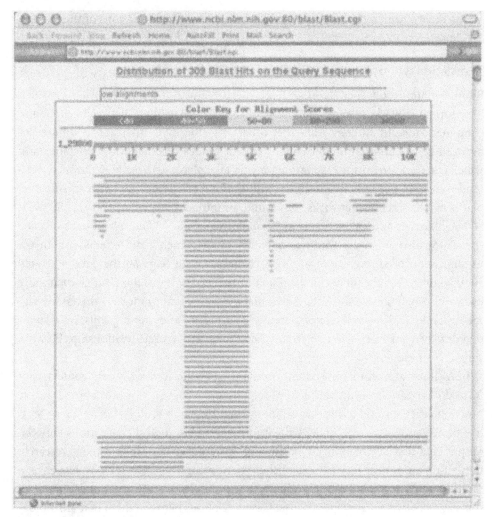

Figure 6. Results of a BLAST search, showing graphical alignments of close matches to a test DNA sequence. Reproduced courtesy of the National Council on Biotechnology Information (NCBI).

When the analysis is finished, the output data is collected, ordered, and formatted (as a plain-text e-mail or as HTML) (Figure 6).

BLAST queries involve an incorporation of "raw" biological data that is ported through the medium (in this case, computer network code), so that it can be processed in a medium-specific context in which it "makes sense." Included with the original input sequence are multiple alignments and the sequence, structural, and bibliographical data associated with those sequences. The output is more than mere bits or letters; the output is a configured bio-logic. From a philosophical-technical perspective, BLAST is not so much a sequence alignment tool, but rather an exemplary case of the ways in which bioinformatics translates a bio-logic across media, with an emphasis on the ways in which the control principle may extend the dimensions of the biological data.

Translation without Transformation

Bioinformatics tools such as BLAST are specific illustrations of the ways in which bioinformatics works toward a condition of "total translatability" between genetic and computer codes. In different ways, they constantly attempt to manage the "difference" between genetic and computer codes through modes of regulating the contexts in which code conversions across varying media take place. As already mentioned, such modes include the mostly automated translations of code types between software, platforms, and across networks (DNA and amino acid sequence; ESTs, SLS, SNPs, cDNA libraries, and other forms of biomolecular code). These techniques of code translation also imbricate another regulatory mode, which is noise reduction, or the minimization of ambiguity regarding a particular code (e.g., all the data generated about a given DNA sequence in GenBank). An effective translation, made even more concise through noise reduction, means that a final regulatory mode—that of managing unstable media—can be implemented as a means of designing software systems that display regular behavior, deterministic actions, and discrete relationships between software components and networks.

BLAST also aims for a transparent translation of code across different contexts, but instead of focusing on universal rules for doing so, BLAST is more focused on the functionalities specific to itself as a bioinformatics search tool. Database query tools like BLAST are primarily concerned with providing bio-logical translations between output data that can be utilized for analysis and further molecular biology research. Presenting an output file that establishes relationships between a DNA sequence, a multiple pairwise alignment, and other related biomolecular information is a mode of translating between data types by establishing relations between them that are biological relations. The way in which this is done is, of course, via a strictly "nonbiological" operation, which is the correlation of data as data (as string manipulation, as in pairwise alignments). In order to enable this technical and biological translation, BLAST must operate as a search tool in a highly articulated manner. A "structured query language" in this case means much more than simply looking up a journal author, title of a book, or subject heading. It implies a whole epistemology, from the molecular biological research point of view. What can be known and what kinds of queries can take place are intimately connected with tools such as BLAST. For this reason tools such as BLAST are excellent as "gene finders," but they are poor at searching and analyzing nested, highly distributed biopathways in a cell, for instance. The principle of unstable media is here stabilized by aligning the functionalities of the materiality of the medium (in this case, computer database query software) with the organization of biological data along two lines (in a database, and as data separated by biological components and processes—DNA, RNA, ESTs, amino acids, etc.).

This look at BLAST points to an important aspect of the practical and theoretical dimensions of "total translatability." In the example of BLAST, the interplay of genetic and computer codes is worked upon so that their relationship to each other is transparent. In other words, when looked at as "biomedia"—that is, as the technical recontextualiztion of the biological domain—BLAST reinforces the notion of media as transparent. The workings of a BLAST search query disappear into the "back end," so that the bio-logic of DNA sequence and structure can be brought forth as genetic data in itself. However, as we have seen, this requires a fair amount of work in recontextualizing and reconfiguring the relations between genetic and computer codes; the regulatory practices pertaining to fluid translation, noise reduction, and the monitoring of unstable media, all work toward this end.

The amount of recontextualizing that is required, the amount of technical reconfiguration needed, is therefore proportionately related to the way in which the bio-logic of DNA sequence and structure is presented. In a BLAST query and analysis, the structure and functionalities of computer technology are incorporated as part of the "ontology" of the software, and the main reason this happens is so that novel types of bio-logic can become self-apparent, characteristics that are "biological" but that could also never occur in the wet lab (e.g., multiple pairwise alignments, gene prediction, cross-database queries). The biological, biomolecular body never ceases to bring itself forward as biological; and this is only made possible by the complex of contexts in which genetic and computer codes are brought into relationships in which a total translatability may take place.

This generates a number of tensions in how the biomolecular body is reconceptualized in bioinformatics. A primary tension is in how translation without transformation becomes instrumentalized in bioinformatics practices. On the one hand, a majority of bioinformatics software tools make use of computer and networking technology to extend the practices of the molecular biology lab. In this, the implication is that computer code transparently brings forth the biological body in novel contexts, thereby producing novel means of extracting data from the body. Indeed, one of the great advantages of moving the human genome effort on-line has been this assumption regarding the transparency of biomedia. The incredible amount of technical infrastructure involved—from automated gene sequencing computers to any number of database, client-server, and security applications—alone testifies to the "high-tech" character of biotech and genetics. Within this infrastructure is also a statement about the ability of computer technology to bring forth, in ever greater detail, the bio-logic inherent in the genome, just as it is in the biological cell. The code can be changed, and the body remains the same; the same bio-logic of a DNA sequence in a cell in a petri dish is conserved in an on-line genome database.

On the other hand, there is a great difference in the fact that bioinformatics does not simply reproduce the wet biology lab as a kind of simulation. Rather, the extension of biology from the wet lab to the dry lab also means, from the technological point of view, that new techniques, new objects of study, and new tools are developed in accordance with the affordances of this particular medium. Techniques such as gene predictions, database comparisons, and multiple sequence analysis generate biomolecular bodies that are specific to the medium of the computer.[47] In this sense, the techniques of bioinformatics imply a certain progressiveness in regard to the instrumentalization of the biomolecular body. The newfound ability to perform string manipulations, database queries, and modeling of data, and to standardize markup languages, also means that the question of "what a body can do" is extended in ways specific to the medium. When we consider bioinformatics practices that directly relate (back) to the wet lab (e.g., rational drug design, primer design, genetic diagnostics), this instrumentalization of the biomolecular body becomes rematerialized in significant ways. Change the code, and you change the body. A change in the coding of a DNA sequence in an on-line database can directly affect the synthesis of novel compounds in the lab.

As a way of elaborating on these tensions, we can further discuss the situation of "translation without transformation" along four main axes.

DNA = data

The first formulation is that of equivalency between genetic and computer codes. As mentioned earlier, the basis for this formulation has a history that extends back to the postwar period, in which the discourses of molecular biology and cybernetics intersect, culminating in what Francis Crick termed "the coding problem," or how DNA, the genetic "code," produced a wide array of proteins.[48] The suggestive tropes and informatic metaphors have been extended in bioinformatics, and materialized in practices such as gene sequencing, and objects such as genome databases. In particular, gene sequencing offers a paradigmatic example of how bioinformatics technically establishes a condition of equivalency between genetic and computer codes. If, so the logic goes, there is a "code" inherent in DNA—that is, a pattern of relationships that is more than DNA's components or substance—then it follows that that code can be identified, "read," and isolated across different media.

A gene sequencing approach, such as Celera's "whole genome shotgun sequencing" method, applies this logic to DNA as both a biological molecule and a digital database.[49] Shotgun sequencing, as its name implies, begins with a genome sample (amplified by methods such as PCR), that is literally blown apart into a large number of smaller fragments. Each of those fragments is "tagged" with a short strand of DNA whose unique sequence is known (called "sequence tagged sites" or STS). Because the tags at the ends of each fragment are known, the pieces of the DNA can then be arranged end to end, so that their sequence is eventually apparent, literally "spelled out" by the binding of the tagged sites. This, however requires a great deal of iterative and combinatorial work, somewhat like piecing together a three-dimensional jigsaw puzzle. For this reason, genome sequencing computers are used, which help to automate the shotgun process.

Although the primary goal of this procedure is to sequence an unknown sample, what shotgun sequencing also accomplishes is the conditions through which an equivalency can be established between the wet, sample DNA, and the reassembled genome sequence that is output by the computer; that is, before any of the more sophisticated techniques of bioinformatics or genomics or proteomics can be carried out, the principles for the technical equivalency between genetic and computer codes must be articulated.

DNA ↔ data

Building on this, the second formulation is that of a back-and-forth mobility between genetic and computer codes; that is, once the parameters for an equivalency can be established, the conditions are set for enabling translations between genetic and computer codes, or for facilitating their transport across different media. The mobility between genetic and computer codes is, then, based on something other than their substance, composition, or components. If this were not the case, there would seem to be an incommensurable difference between biological molecules and silicon integrated-circuit (IC) chips, between the biochemistry of DNA and the algorithmic logic of computer software. However, bioinformatics practices, while recognizing this difference, also privilege another view of the relation between genetic and computer codes, and that is one based on identified patterns of relationships between components—for instance, the bio-logic of DNA's base pair binding scheme (A-T; C-G), or the identification of certain polypeptide chains with folding behaviors (amino acid chains that make alpha-helical folds). These well-documented characteristics of the biomolecular body become, for bioinformatics, more than something defined by its substance. They are configured by the conceptual and practical aspects of bioinformatics, as a codifiable set of relationships, as a pattern of relationships that, as a pattern, may exist through a variety of material substrates or media.

The mobility between genetic and computer codes is more than the mere digitization of the former by the latter. It is a set of techniques that build on the parameters set by the software, which establish a beginning equivalency between genetic and computer codes. The mobility between them is the extension of specified patterns of relationships (e.g., base pair binding, transcription, translation, protein folding) beyond the material substrate of the biological domain (e.g., cells, chromosomes, plasmids, clone libraries). What makes DNA in a plasmid and DNA in a database the "same"? A definition of what constitutes the "biological" in the term *biological data*. What computer science brings to the concepts of molecular biology is the same thing that genetic engineering techniques have brought to it: namely, a view of "the biological" as more than static substance, but relations and functionalities that can be isolated, extracted, relocated, and designed in a range of experimental contexts.

Once "the biological" is defined not as substance but as patterns of relationships, then the equivalency between genetic and computer codes can take on a more pragmatic tone. Once the biological domain is approached as these identified and selected patterns, then, from one perspective—the perspective of bioinformatics—it matters little whether that DNA sample resides in a test tube or a database. An example is any database query for a nucleotide or amino acid sequence. Although, in purely computational terms, this is nothing more than pattern matching strings of data, in bioinformatic terms it is part of a process in which the bio-logic of DNA or protein code is conserved in the database—or, we should say, through the database. The mobility between genetic and computer codes not only means that the "essential data" or patterns can be translated from wet to dry DNA. It also means that bioinformatics practices such as database queries are simultaneously algorithmic (following computer logic) and biological (following the conserved pattern of relationships in the computer, the bio-logic).

DNA } data

Beyond these two first formulations are others that develop functionalities based on them. One is that in which data accounts for the body. In examples relating to genetic diagnostics, data does not displace or replace the body, but rather forms a kind of index to the informatic muteness of the biological body—DNA chips in medical genetics, disease profiling, preimplantation screening for in vitro fertilization (IVF), as well as other, nonbiological uses (DNA fingerprinting in law, genetic ID tags in the military). The examples of DNA chips in medical contexts redefine the ways in which accountability takes place in relation to the physically present, embodied subject. Although its use in medicine is far from being common, genetic testing and the use of DNA chips are, at the very least in concept, integrating themselves into the fabric of medicine, where an overall genetic model of disease is often the dominant approach to a range of conditions, from Alzheimer's to diabetes to cancer.[50]

Although genetic testing and the use of DNA chips are highly probabilistic and not, by any means, deterministic, the way in which they configure the relationship between genetic and computer codes is likely to have a significant impact in medicine. Genetic testing can, in best-scenario cases, tell a patient the likelihood of potentially developing conditions in which a particular disease may or may not manifest itself, given the variable influences of environment and patient health and lifestyle. In a significant number of cases, this amounts to "leaving things to chance," and one of the biggest issues with genetic testing is the "decision to know."

However, beneath these issues is another set of questions, which pertain to the ways in which a mixture of genetic and computer codes gives testimony to the body, and through its data output, accounts for the body in medical terms (genetic patterns, identifiable disease genes, "disease predispositions"). Again, as with the establishing of an equivalency, and the effecting of a mobility, the complex of genetic and computer codes must always remain biological, even though its very existence is in part materialized through the informatic protocols of computer technology. The biomolecular body in medical genetics is recontextualized as a body that "scripts," in the sense of scripting languages in computer programming, those quasi languages that often operate on top of full-blown languages (e.g., Javascript, Perl, PHP). The scripts point back to a site within the biological body, in the way that a Javascript will point to a site within the main body of an HTML document. This body that scripts is a collection of output data and a technical apparatus, such as DNA chip, analysis computer, and diagnostic software. Thus, in a situation where data accounts for the body, we can also say that a complex of genetic and computer codes makes use of the mobility between genetic and computer codes, to script an indexical description of the biological domain. But, it should be reiterated that this scripted data is not just data, but a conservation of a bio-logic, a pattern of relationships carried over from the patient's body to the DNA chip to the computer system. It is in this sense not only that data accounts for the body, but that the data (a complex of genetic and computer codes) also identifies itself as biological.

data } DNA

Finally, not only can data account for the body, but the body can be generated through data. Although this notion would seem to belong more to the domain of science fiction, we can see this type of formulation occurring in practices that make use of biological data to effectuate changes in the wet lab, or even in the patient's body. Here we can cite the field known as "pharmacogenomics," the use of genomic data to design custom-tailored gene-based and molecular therapies.[51] Pharmacogenomics moves beyond the synthesis of drugs in the lab, for several reasons. It relies on data from genomics in its design of novel compounds, and in this sense makes use of simulation software, such as protein docking applications. As a business model, it is also linked to the use of DNA chips and other medical diagnostic devices, in that these provide the means by which a patient's genome can be constructively compared to "the" genome, thereby making possible the identification of polymorphisms, point mutations, and other identifiable anomalies. Pharmacogenomics is based not on the diagnostic model of traditional pharmaceuticals (ameliorating symptoms), but rather on the model of "preventive medicine" (using predictive methods and genetic testing to prevent disease occurrence to begin with). This means not only that pharmacogenomic therapies will be custom-designed to each individual patient's genome, but that the therapies will integrate themselves into the biomolecular body of the patient in the long term.

What this means is that "drugs" are replaced by "therapies," and the synthetic is replaced by the biological, but a biological that is not only curative, but preventive as well. The image of the immune system that this evokes is one based on more than the correction of "error" in the body. Rather, it is based on the principles of biomolecular and biochemical design, as applied to the optimization of the biomolecular body of the patient. At an even more specific level, what is really being "treated" is not so much the patient as the genome. At the core of pharmacogenomics is the notion that a reprogramming of the "software" of the genome is the best route to preventing the potential onset of disease. If the traditional approaches of immunology and the use of vaccines are based on synthesizing compounds to counter certain proteins (e.g., compounds to bind to cell surface recognition sites on a pathogen), the approach of pharmacogenomics is to create a context in which a reprogramming (or "recoding") will provide an instance in which the body will biologically produce the needed antibodies itself. The aim of pharmacogenomics is, in a sense, to not make any drugs at all, but to enable the patient's own genome to do so.[52]

In order for this to occur, pharmacogenomics relies on bioinformatics to query, analyze, and run comparisons against genome and proteome databases. The data gathered from this work is what sets the parameters of the context in which the patient genome may undergo gene-based, cell, or molecular therapy. The procedure may be as concise as prior gene therapy human clinical trials—the insertion of a needed gene into a bacterial plasmid, which is then injected into the patient. Or, the procedure may be as complex as the introduction of a number of inhibitors that will collectively act at different locations along a chromosome, working to effectuate a desired pattern of gene expression to promote or inhibit the synthesis of certain proteins. In either instance, the dry data of the genome database extends outward and directly rubs up against the wet data of the patient's cells, molecules, and genome. In this sense, data generates the body, or, the complex of genetic and computer codes establishes a context for recoding the biological domain.

The equivalency, back-and-forth mobility, and the accountability and generativity of code in relation to the body are all ways in which the biomolecular body as "biomedia" is achieved. As we have already seen, the prerequisite for biomedia is this tension: conserving bio-logics across media, while also instantiating modes of reconfiguring bio-logics by reconfiguring code. This situation, of translation without transformation, is not, however, determined by the medium of the computer; nor is it inherent to processes within the digital domain. There are, we can suggest, many relationships between genetic and computer codes, in which biomedia are not simply moments in which the biological is interpreted as a transparent medium.

Virtual Biologies

Thus far, we have seen how bioinformatics is much more than the simple application of computer technology to bioscience research. In viewing bioinformatics as a set of practices that attempt to regulate the "difference" between genetic and computer codes, we have seen how a "bio-logic" is simultaneously conserved and modulated or recoded across different media. We might abbreviate by saying that, in relation to the biomolecular body, bioinformatics aims for a cross-platform compatibility.

However, in this regulation of genetic and computer codes, we have also seen that bioinformatics can be described as "translation without transformation." The result of this is that the manifold relationships between DNA and data are given forth as both transparent and reprogrammable. Because the data from a genome database is not static, but constantly encoded, recoded, and decoded, there is an ambivalence as to whether that data is in some way "DNA itself," or merely an informatic representation of a "thing." When this biological data is ported through the gene prediction, drug targeting, and clinical trial pipelines, DNA is taken not as a "thing" but as selected patterns of relationships that remain, be it on a computer screen or in an in vitro cell culture. But, at the same time, biological data has no functional autonomous existence separate from the referent of the "wet" biological domain. This is a referent that selectively reinforces the commonsensical notion that bioinformatics is merely simulation, whereas cell cultures in a wet lab are working with "DNA itself."

One of the primary challenges for bioinformatics is to rethink this dichotomy as "biomedia," or the technical recontextualization of the biological domain. As we have seen, "biomedia" is first and foremost a critical concept: it points to the assumptions inherent in fields such as bioinformatics as regards the division between biology and technology, living and nonliving, organism and machine. But it also points to the many interstitial sites and fissures that exist in these dichotomous relationships (a "technology" that implies the biological as biological). Perhaps, if there is one thing to be gleaned from the concept of "biomedia," it is that, in biotech research, what counts as biological and/or living cannot be dissociated from the technical and instrumental means of articulating the biological. In the case of bioinformatics, this means reconsidering the tensions that exist between the two types of information, genetic and computer codes. If, as I have suggested, bioinformatics operates via a "bio-logic" of "translation without transformation," we can also ask how the contradictions between genetic and computer codes alluded to earlier may be addressed in a way that would transform our assumptions concerning the biology–technology division.

As a suggestion, we can consider how various "open-source" movements have contributed to bioinformatics.[53] In its more specific guise, open source can be described as a set of programming and reprogramming practices centered on a given technical community. Linux, for instance, has several open-source communities (Red Hat, Yellow Dog, GNU Linux, linuxppc), each of which is united by a common set of interests that are, first and foremost, technical. Open-source communities often work by means of a download-tweak-upload procedure, in which a single piece of code may be authored by multiple people, and will have gone through several versions. Theoretically, with open source nothing is proprietary, and the emphasis is generally on community-based development of software tools. Open-source initiatives must therefore toggle a line between calcified standardizations, on the one hand, and absolute diversity, on the other.

Although early bioinformatics examples began at universities, bioinformatics is quickly becoming a not insignificant part of the software industry. Commercial bioinformatics suites, which perform multiple types of analyses and searches, can easily cost thousands of dollars, and often require specialized access to company-owned proprietary databases, as well as an extra charge for customer support. The molecular biologist today would seem to have two basic choices for doing computational research: either university- or government-subsidized tools (such as BLAST or GenBank), or corporate packages (such as those by Incyte Genomics, eBioinformatics, or Perkin-Elmer).

In the 1990s, a third option emerged, one that posed a challenge to the bioinformatics industry, and that made use of the free university-based projects. "Opensource bioinformatics" (or simply OSB) describes this third option. As with other open-source movements, OSB is geared toward the multiple-authored and -versioned community-based development of software tools for bioinformatics. A case in point is the bioperl.org group. Bioperl is based on the Perl programming language, one often used in handling data strings (such as text or numbers, but also DNA or protein code), and also in text processing in client–server interactions (such as CGI).[54]

Perl has, from its beginning, served as an example of a programming language geared toward open sourcing and a high degree of customization. In fact, a common phrase found in Perl circles is "there's more than one way to do it" (TMTOWTDI, sometimes pronounced "tim toady"). The BioPerl project's use of Perl begins around 1996, when Lincoln Stein, a programmer and molecular biologist at the Whitehead Institute, developed specialized Perl modules for genome sequencing projects then under way. Because genome sequencing involved handling large amounts of text data (DNA or protein code), Perl was used to help automate and facilitate the organization and analysis of biological data.[55]

Just as bioinformatics tools such as BLAST operate on the control principle of the "query," the BioPerl project works through a modular principle based on the notion of "source code." For instance, a section of a Perl module for sequence alignment of DNA or protein code looks as follows:

```
use Bio::Seq;
use Bio::SimpleAlign;
use Bio::AlignIO;

$tseq = 'SKSESPKEPEQLRKLFIGGLSFETTDESLRSHFEQWGTLTDCVVMRDPNTKRSRGFGFVT
YATVEEVDAAMNARPHKVDGRVVEPKRAVSREDSQRPGAHLTVKKIFVGGIKEDTEEHHL
RDYFEQYGKIEVIEIMTDRGSGKKRGFAFVTFDDHDSVDKIVIQKYHTVNGHNCEVRKAL
SKQEMASASSSQRGRSGSGNFGGGRGGGFGGNDNFGRGGNFSGRGGFGGSRGGGGYGGSG
DGYNGFGNDGGYGGGGPGYSGGSRGYGSGGQGYGNQGSGYGGSGSYDSYNNGGGRGFGGG
SGSNFGGGGSYNDFGNYNNQSSNFGPMKGGNFGGRSSGPYGGGGQYFAKPRNQGGYGGSS
SSSSYGSGRRF';

$tseq = ~s/[^A-Z]//g;

$seq1 = Bio::Seq->new(-id=>'roa1_human',-seq=>$tseq);

$tseq = 'MVNSNQNQNGNSNGHDDDFPQDSITEPEHMRKLFIGGLDYRTTDENLKAHFEKWGNIVDV
VVMKDPRTKRSRGFGFITYSHSSMIDEAQKSRPHKIDGRVVEPKRAVPRQDIDSPNAGAT
VKKLFVGALKDDHDEQSIRDYFQHFGNIVDINIVIDKETGKKRGFAFVEFDDYDPVDKVV
LQKQHQLNGKMVDVKKALPKQNDQQGGGGGRGGPGGRAGGNRGNMGGGNYGNQNGGGNWN
NGGNNWGNNRGGNDNWGNNSFGGGGGGGGGYGGGNNSWGNNNPWDNGNGGGNFGGGGNNW
NNGGNDFGGYQQNYGGGPQRGGGNFNNNRMQPYQGGGGFKAGGGNQGNYGGNNQGFNNGG
NNRRY';
```

```
$tseq = ~s/[^A-Z]//g;

$seq2 = Bio::Seq->new(-id=>'roa1_drome',-seq=>$tseq);

$tseq = 'MHKSEAPNEPEQLRKLFIGGLSFETTDESLREHFEQWGTLTDCVVMRDPNSKRSRGFGFV
TYLSTDEVDAAMTARPHKVDGRVVEPKRAVSREDSSRPGAHLTVKKIFVGGIKEDTEEDH
LREYFEQYGKIEVIEIMTDRGSGKKRGFAFVTFEDHDSVDKIVIQKYHTVNNHNSQVRKA
LSKQEMASVSGSQRERGGSGNYGSRGGFGNDNFGGRGGNFGGNRGGGGGFGNRGYGGDGY
NGDGQLWWQPSLLGWNRGYGAGQGGGYGAGQGGGYGGGQGGGYGGNGGYDGYNGGGSGF
SGSGGNFGSSSGGYNDFGNYNSQSSSNFGPMKGGNYGGGRNSGPYGGGYGGGSASSSSGYG
GGRRF';

$tseq = ~s/[^A-Z]//g;
$seq3 = Bio::Seq->new(-id=>'roa1_xenla',-seq=>$tseq);
```

Perl modules are primarily composed of "functions" and "statements." Functions are commands or actions performed in the Perl environment. In the case of BioPerl, functions may include retrieving a sequence from a database, aligning sequences, or editing sequences. These are specific to molecular biology, but the actual functions are ones common to Perl programming (e.g., the "loop" function may be used iteratively to align a sequence letter by letter). Statements are often contained within functions, as the parameters of a function. What those parameters actually are in the BioPerl example are statements that may relate to the type of data (nucleotide or amino acid) or database tool being accessed (Pfam, Prosite). Beyond this, what the Perl module operates on are the numbers, strings, and variables specific to the bioinformatic view of the biological domain (numbers that relate to probabilities in an alignment, a string of sequence text, or variables denoting a gene ID tag). Even without explaining every aspect of the code above, what is immediately evident is that it involves a particular combination of genetic and computer codes. In fact, the module was written specifically to handle the ways in which the bio-logic of amino acid sequence informatically "aligns" itself in the cell (informatically, not structurally, note).

Perl modules are often short, and designed for modularity, enabling the programmer to take a piece of code here, and some there, and then develop a specialized module for a particular task. In the case of BioPerl, this involves combining data-processing capabilities (manipulating strings) with the particular way in which biological data is processed in the cell; that is, BioPerl is based not only on the conservation of a bio-logic across media, but on the conservation of the way in which DNA itself procesess data. In this sense, BioPerl's meaning of source code does not mean "origin" or "reference," but rather generates a context, an environment, in which biological data generates further extensions of biological data.

The BioPerl modules themselves do not contain a great deal of code; the abbreviated character of Perl's functions, statements, and variables makes for code that is made up less of instructions, and more of designs for contexts. Perl is not a product of bioinformatics research; many of the "universal" functions in Perl (such as the "print" function, which displays output data) are regularly used in the BioPerl modules. It is, rather, the way in which such functions are used that makes BioPerl more than simply another, purely computational implementation of the Perl programming language. In both the way in which Perl code contextualizes biological data and the way in which Perl code is generated, "open source" implies a transformation of code. Someone developing a specific Perl module downloads examples, tweaks them, recodes them, then implements it and uploads the new version. This open-source bioinformatics is, then, not only "open" to novel technical developments, but, being open source, is also highly flexible and adaptive to different needs of the technical community. In a sense, an OSB project like BioPerl is a body of code with a morphology, one that is constantly being tweaked and modulated.

The main question, however, is whether this transformation of code can also be a transformation of practice, or, more fundamentally, a transformation of how questions are asked in bioinformatics practices. As we have seen, the intersection of biotech and infotech goes by many names—bioinformatics, computational biology, virtual biology. This last term is especially noteworthy, for it is indicative of the kinds of philosophical assumptions within bioinformatics that we have been querying. A question, then: is biology "virtual"? Certainly, from the perspective of the computer industry, biology is indeed virtual, in the sense of "virtual reality," a computer-generated space within which the work of biology may be continued, extended, and simulated. Tools such as BLAST, molecular modeling software, and genome sequencing computers are examples of this emerging virtual biology. From this perspective, a great deal of biotech research—most notably the various genome efforts—is thoroughly virtual, meaning that it has become increasingly dependent on and integrated with computing technologies.

But, if we ask the question again, this time from the philosophical standpoint, the question changes. Asking whether or not biology is philosophically virtual entails a consideration of how specific fields such as bioinformatics conceptualize their objects of study in relation to processes of change, difference, and transformation. If bioinformatics aims for the technical condition (with ontological implications) of "translation without transformation," then what is meant by "transformation"? As we have seen, transformation is related technically to the procedures of encoding, recoding, and decoding genetic information that constitute a bio-logic. What makes this possible technically is a twofold conceptual articulation: that there is some thing in both genetic and computer codes that enables their equivalency and therefore their back-and-forth mobility (DNA sampling, analysis, databasing). This technical-conceptual articulation further enables the instrumentalization of genetic and computer codes as being mutually accountable (genetic disease predisposition profiling) and potentially generative or productive (genetically based drug design or gene therapies).

Thus, the transformation in this scenario, whose negation forms the measure of success for bioinformatics, is related to a certain notion of change and difference. To use Henri Bergson's distinction, the prevention of transformation in bioinformatics is the prevention of a difference that is characterized as being quantitative (or "numeri-cal") and extensive (or spatialized).[56] What bioinformatics developers want to prevent is any difference (distortion, error, noise) between what is deemed to be information as one moves from an in vitro sample, to a computer database, to a human clinical trial for a gene-based therapy. This means preserving information as a quantifiable, static unit (DNA, RNA, protein code) across variable media and material substrates.

However, difference in this sense—numerical, extensive difference—is not the only kind of difference. Bergson also points to a difference that is, by contrast, qualitative ("nonnumerical") and intensive (grounded in the transformative dynamics of tempo-ralization, or "duration"). Gilles Deleuze has elaborated Bergson's distinction by refer-ring to the two differences as external and internal differences, and has emphasized the capacity of the second, qualitative, intensive difference to continually generate difference internally—a difference from itself, through itself.[57]

How would such an internal—perhaps self-organizing—difference occur? A key concept in understanding the two kinds of differences is the notion of "the virtual," but taken in its philosophical and not technical sense. For Bergson (and Deleuze), the virtual and actual form one pair, contrasted to the pair of the possible and the real. The virtual/actual is not the converse of the possible/real; they are two different processes by which material-energetic systems are organized.[58] The possible is negated by the real (what is real is no longer possible because it is real), and the virtual endures in the actual (what is actual is not predetermined in the virtual, but the virtual as a process is immanent to the actual). As Deleuze notes, the possible is that which manages the first type of difference, through resemblance and limitation (out of a certain number of possible situations, one is realized). By contrast, the virtual is itself this second type of difference, operating through divergence and proliferation.

With this in mind, it would appear that bioinformatics—as a technical and con-ceptual management of the material and informatic orders—prevents one type of difference (as possible transformation) from being realized. This difference is couched in terms derived from information theory and computer science, and is thus weighted toward a more quantitative, measurable notion of information (the first type of quan-titative, extensive difference). But does bioinformatics—as well as molecular genetics and biology—also prevent the second type of qualitative, intensive difference? In a sense it does not, because any analysis of qualitative changes in biological information must always be preceded in bioinformatics by an analysis of quantitative changes, just as genotype may be taken as causally prior to phenotype in molecular genetics. But, in another sense, the question cannot be asked, for before we inquire into whether or not bioinformatics includes this second type of difference in its aims (of translation with-out transformation), we must ask whether or not such a notion of qualitative, inten-sive difference exists in bioinformatics to begin with.

This is why we might question again the notion of a "virtual biology"; for, though bioinformatics has been developing at a rapid rate in the past five to ten years (in part bolstered by advances in computer technology), it still faces a number of extremely difficult challenges in biotech research. Many of these challenges have to do with biological regulation: cell metabolism, gene expression, and intra- and intercellular signaling.[59] Such areas of research require more than discrete databases of sequence data; they require thinking in terms of distributed networks of processes that, in many cases, may change over time (gene expression, cell signaling, and point mutations are examples).

In its current state, bioinformatics is predominantly geared toward the study of discrete, quantifiable systems that enable the identification of something called genetic information (via the fourfold process of bio-logic). In this sense, bioinformatics works against the intervention of one type of difference, a notion of difference that is closely aligned to the traditions of information theory and cybernetics. But, as Bergson reminds us, there is also a second type of difference, which, while being amenable to quantitative analysis, is equally qualitative (its changes are not of degree, but of kind) and intensive (in time, as opposed to the extensive in space). It would be difficult to find this second kind of difference within bioinformatics as it currently stands; however, many of the challenges facing bioinformatics—and biotech generally—imply the kinds of transformations and dynamics embodied in this Bergsonian–Deleuzian notion of difference-as-virtual.

It is in this sense that a "virtual biology" is not a conceptual impossibility, given certain contingencies. The "data" of bioinformatics establishes networks between biological samples in the wet lab, biological databases, software and programming languages, prediction and design of biomolecules, and, in the end, a further rematerialization in vitro (genomics and proteomics research) or in vivo (genetic medicine, drug targets, clinical trials). In no two cases does "biological data" mean exactly the same thing, even though a "bio-logic" (patterns of relationships) is conserved in each case. If bioinformatics is to accommodate the challenges put to it by the biological processes of regulation (metabolism, gene expression, signaling), then it will have to consider whether or not a significant reworking of what counts as "biological information" will become necessary. As has been pointed out, this reconsideration of information will have to take place on at least two fronts: that of the assumptions concerning the division between the material and informatic orders (genetic and computer codes, biology and technology, etc.), and that of the assumptions concerning material-informatic orders as having prior existence in space, and secondary existence in duration (molecules first, then interactions; objects first, then relations; matter first, then force).[60] The biophilosophy of Bergson (and Deleuze's reading of Bergson) serves as a reminder that, although contemporary biology and biotech are incorporating advanced computing technologies as part of their research, this still does not necessarily mean that the informatic is "virtual."

CHAPTER THREE

Wet Data

Biochips and BioMEMS

Blood, Biochips, and Rhythms

In Greg Bear's 1985 novel *Blood Music,* Virgil Ulam, a nerdy biochemist working on engineered protein-silicon biochips, takes his research to the extremes of self-experimentation, inserting the "intelligent" biochips into his white blood cells. The biochips begin to integrate themselves into his biological system, improving everything from his metabolism to his skeletal structure, even his sexual performance. As Ulam confides to a friend, "I'm being rebuilt from the inside out."[1]

However, the biochips' ability to adapt and improve their own complexity lead to horrifying results. Ulam continues, "Then one night my skin started to crawl ... I wondered what they'd do when they crossed the blood-brain barrier and found out about me—about the brain's real function."[2]

In the biochips' programmed endeavor to improve and complexify themselves, Ulam's body shifts from being an improved human anatomy to something very nonhuman. As the narrator puts it: "In hours, our legs expanded and spread out. Then extensions grew to the windows to take sunlight, and to the kitchen to take water from the sink. Filaments soon reached to all corners of the room ... I suspect we resemble cells."[3] As the novel progresses, the biochips gain a collective intelligence, forming "noocytes" that eventually incorporate not only bodies but also the natural environment and entire cities. The story closes with a giant question mark as to the fate of "the human" as it is subjected to highly sophisticated biomolecular technologies.

Blood Music shows us the ways in which biotechnologies become "biomedia": a technology for the mediation of the biological, across different material substrates, and spanning the macro- and microscale worlds. What begins as tools for medicine and pharmaceuticals—in *Blood Music* the protein-silicon lymphocytes for diagnostics—ends up becoming a proliferator of biological complexity, a new type of biotechnically "intelligent" epidemic. The "noocytes" or intelligent cells displace the organization of the central nervous system as a mode of operating the body. They form a set of nodes or cell clusters in a highly decentralized manner, transgressing bodily boundaries, body–environment boundaries, and the boundaries separating organic from non-organic, living from nonliving. Able to "communicate" using proteins and enzymes, the noocyte clusters literally efface the outlines of the human body: tendrils reach out from amorphous bodies, bony ridges or communication channels form on the surface of the skin, and the skin itself, the membrane defining the anthropomorphic body, melds with other organic matter, in a kind of macroscale molecular "culture." The result, as Bear presents it, is a moment of both apocalypse and transformation, though the exact nature of the transformation is ambivalent. Behind such transformations is, however, an assumption by the researchers in the novel that DNA is, essentially, information. As one researcher states in the novel:

> What I am saying is that is that we now have conditions sufficient to cause the effects I've described in my papers. Not just four, five billion individual cogitators, Michael, but trillions . . . perhaps billions of trillions . . . Tiny, very dense, focusing their attention on all aspects of their surroundings, from the very small to the very large. Observing everything in their environment, and theorizing about the things they do not observe. Observers and theorizers can fix the shape of events, of reality, in quite significant ways. There is nothing, Michael, but information.[4]

This assertion—very alive in current biotechnology and genetics research—that DNA is information, and that the body is, fundamentally, all information, is a new type of "genetic reductionism" that relies heavily on the ubiquitous properties of data, be it electronic or biological data. Once genetics approaches DNA as information, or more importantly, as a computer, it is a short step to inquiring whether genomes and digital files, DNA and data, or molecules and bits, can be freely exchanged across any medium. Genomic databases, "e-labs," and advanced molecular simulation are all scientific instances of this position that "biology is information."

Blood Music expresses several anxieties surrounding the ways in which our bodies may be transformed by fields such as biotechnology and nanotechnology. On one level, the very idea of technically integrating DNA or proteins with silicon microchips represents an uneasy fusion of bodies and technologies, flesh and machine, a discomfort concerning the unpredictability of these novel hybrids. In addition, these modifications of the body–technology relationship are predicated on the ability to program and control life at the molecular level—something that has been a technical mainstay of molecular biology for some time. Implantable biosensors, DNA chips, and tiny biological micromachines are just some examples of hybrid biotechnologies currently being developed from this informatic view of the body.

Despite their highly technical and pragmatic uses, these tiny devices raise a series of philosophical and technical questions, pertaining to the shifting relationships between the body and technology: How do these microdevices transform the notion of the biological body in biomedicine and biotechnology? Are they simply tools, or are they in some way viable, living systems? Do they prompt us to rethink our common definitions of media and technology? How do they handle the different "data types" that flow across biological and nonbiological media, and how to those data types relate to the biomolecular body of a genome or an individual patient?

Here we will consider such hybrid "microsystems" as material-informatic instances in which the boundary between biology and technology is constantly negotiated. In cases where such devices are regularly used as part of laboratory research for genetic screening, or cases in which they are implanted into the body of patients for in vivo biomonitoring, this boundary negotiation between the biological and technological, the molecular and the informatic, becomes an explicitly social and political one as well. Although devices such as DNA chips would seem to be exemplary instances of the "cyborg," we can suggest that something more complex is at issue, a particular technical mediation of the biological domain that we have been calling "biomedia."

Microsystems

The dizzying array of new technologies, techniques, and devices emerging out of biotechnology and its related fields can obfuscate the fact that such devices form distinct classes of technologies, each with their own perspective on the relationship between biology and technology, on the limits to the technical uses of life, or indeed on the very issue itself of what constitutes "life." Broadly speaking, we can refer to this array of devices, gadgets, and techniques in biotechnology as "microsystems." They share the common characteristic of bringing the biological and technological materials into complex and nuanced relationships—relationships that often frustrate a reduction into the oppositions of nature/culture, or body/technology. Often, as we shall see, the defining characteristic of a particular microsystem is not what it is made of (its components), but rather how it operates, what it does (its organization). As a way of analyzing research into hybrid biotechnology devices, we can distinguish between three general types of research, based both on the materialities of the objects in question and on their contextualization (relationships, applications, functions).

The first group we can call "bioMEMS," and they include engineered hybridizations between the organic and inorganic domain, between cells and integrated circuits, between DNA and silicon.[5] The "MEMS" in bioMEMS stands for "microelectromechanical systems," with the "bio" prefix referring to their intended use in medical and diagnostic contexts. Most often these literal fusions of DNA and silicon are used in biomedical engineering (such as implantable biosensors), in laboratory analysis of biological materials (such as genomes or proteomes), or in neural net or AI-based research involving hybrid cell-circuit fusions (neurons communicating with IC chips). The general goal of this research is to fuse the affordances of biomolecular activity (e.g., DNA's complementary binding) with the affordances of electronic and micromechanical apparatuses (e.g., silicon adhesion principles and microfabrication techniques).

A second group we can refer to as "biocomputing," and it includes research in nanobiotechnology, or the design and engineering of computing systems at the molecular and atomic level.[6] Such research usually aims either to build a new generation of integrated circuits and computers (nanocomputers) using biomolecules, or to use techniques from nanotechnology to design new classes of biomolecules or biomolecular "assemblers." Unlike the MEMS group, the biocomputing group does not make use of inorganic or electromechanical properties or assemblages. In fact, the entire point of biocomputing research is to do away with the very division between organic/inorganic and biological/technological. However, aside from such theoretical considerations, biocomputing's main argument is that a new generation of computers can emerge from the use of biomolecules (such as DNA) rather than electronic integrated circuits.

Finally, a third group is generally known as "artificial life" or a-life.[7] Whereas biocomputing initiatives aim to develop computers out of biology, a-life aims to develop novel means of modeling, and, in some cases, generating, a biology out of computers. The discourse of a-life has, in the past decade or so, expanded considerably, and an overview of this discourse lies outside the concerns of this essay. However, we can note that a-life research is often poised between simulation and generation; that is, if one of the primary arguments of a-life is that "life" be considered not according to substance (such as carbon-based organisms), but rather according to process (such as evolution and self-replication), then the notion of "life" itself becomes at once more flexible and yet dangerously relativized. The use of techniques such as genetic algorithms in a-life has raised the question of whether computers and computer data can in some way be considered "alive."

For our purposes in this essay, we want to focus on the first group—bioMEMS—as instances in which the relationship between biology and technology is constantly put into play. BioMEMS refer to a group of microsystems devices that explicitly combine biomolecules with nonorganic substrates such as silicon, and which are used primarily for research and diagnostics. As mentioned, "MEMS" stands for "microelectromechanical systems," and the "bio" prefix refers to MEMS technologies applied to biomedicine and biotechnology: biological microelectromechanical systems. One of the earliest MEMS-based research programs was initiated by the U.S. government, at DARPA (Defense Advanced Research Projects Agency). According to one of its textbooks on MEMS:

> Using the fabrication techniques and materials of microelectronics as a basis, MEMS processes construct both mechanical and electrical components. Mechanical components in MEMS, like transistors in microelectronics, have dimensions that are measured in microns . . . MEMS is not about any one single application or device, nor is it defined by a single fabrication process or limited to a few materials. More than anything else, MEMS is a fabrication approach that conveys the advantages of miniaturization, multiple components and microelectronics to the design and construction of integrated electromechanical systems.[8]

To this we can add a comment from MEMS Clearinghouse, an on-line hub for the MEMS industry, which states that "MEMS promise to revolutionize nearly every product category by bringing together silicon-based microelectronics with micromachining technology, thereby making possible the realization of complete systems-on-a-chip."[9]

Although MEMS have found application in everything from air bags in cars to cell phones to digital projection, one of the leading fields of MEMS research has been in biotech and biomedicine.[10] Examples of such bioMEMS currently in development include: in vivo blood-pressure sensors (with wireless telemetry), oligonucleotide microarrays (DNA chips), and microfluidics stations (labs on a chip).[11] BioMEMS include not only devices with electrical and mechanical components, but they also include devices that are produced using microelectronics fabrication technologies.[12] BioMEMS engineering is based on the notion that there is a specific "fit" between the biological and technical parts of the system for its particular purpose; it not only combines biological and nonbiological components, but it does so in a way that is functional—both biologically and technologically speaking.[13]

The Shape of BioMEMS

As a way of further understanding the ways in which bioMEMS place the biological and technological in relation to each other, we can begin with a formal analysis of bioMEMS. There are many different shapes and flavors of bioMEMS, from drug-delivery probes the size of a pill, to multilayered chips for the analysis of biological samples.[14] However, bioMEMS all share the commonalities of bringing together biological and nonbiological materials into particular, engineered configurations, where they function in some integrated relationship to biological systems. There are several classes of real-world applications in bioMEMS, and each provides us with a particular type of relationship between bodies and technologies.

One primary area of application is in medical diagnostics. BioMEMS in this class include in vivo biosensors, for measuring blood pressure, as well as in vivo drug probes and chip-based optical prosthetics.[15] In each case, a functioning MEMS device is implanted in the body, where it aims to function invisibly, as it were, in the biological milieu, while also gathering data about its environment (e.g., blood-pressure levels), and then acting on that data (e.g., releasing a drug compound into the bloodstream). In more long-term scenarios, such bioMEMS would also be able to use wireless telemetry to send data outside the body to a receiving computer system (e.g., physician-based or home monitoring using telemedicine technologies). The main significance of such biosensors lies in the fact that they are designed to function as part of biological systems, be it for therapeutic or diagnostic purposes. BioMEMS cannot function without a biological milieu into which they insert themselves, and out of which they construct a particular biological network, incorporating biomolecules, MEMS devices, and managing data signals. In the context of medical diagnostics, bioMEMS highlight the relationship between a bioMEMS device and the biological milieu, ensuring

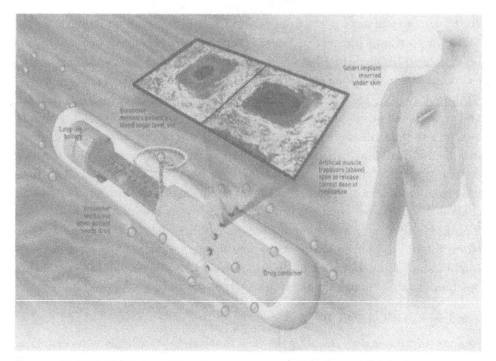

Figure 7. BioMEMS device intended as biosensor to measure blood sugar levels in vivo.
Reproduced courtesy of Dr. Marc Madou.

the right "fit" between them. The body's biological interior thus becomes a system re-
ceptive to a monitoring from the inside out (Figure 7).

   Another area of application includes biological sample preparation for biological
assays or for biological sampling and screening. The microdevices used for these purposes are known as "microfluidics stations" (or "labs on a chip"), because they integrate several sample preparation processes that are conventionally separated to different parts of the molecular biology lab.[16] Most often they are used for laboratory tasks
of isolating, purifying, and amplifying desired components, such as DNA from a
group of cells. A number of companies, such as Nanogen, Caliper, Aclara, and Orchid
Biosciences, are specializing in producing off-the-shelf microfluidics devices for laboratory research. Microfluidics use integrated-circuit manufacturing techniques to
design handheld devices with a number of channels, reservoirs, and basins for directing
the flow of selected biomolecules. The biological samples are processed or organized
using techniques ranging from polarizing the electrical charges of molecules, to the
use of vacuum pressure in channels, to the use of micromechanical pump forces.
Microfluidics stations are otherwise inert devices, until a biological sample is passed
through them—that is, until something "resists" the physics of the device. In the context of biological sample preparation, bioMEMS highlight the relationship between
biological and electromechanical components. Using either electrical conductivity

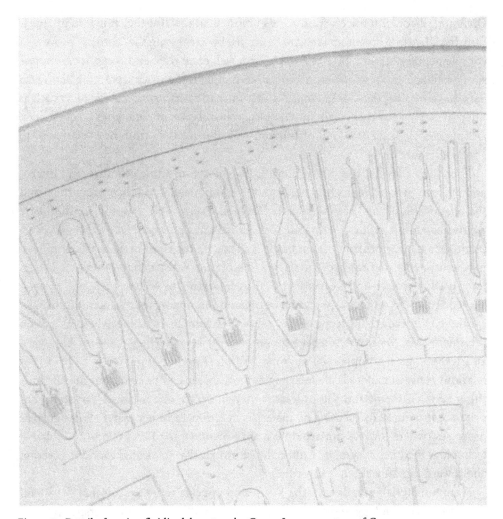

Figure 8. Detail of a microfluidics laboratory by Gyros. Image courtesy of Gyros.

or mechanical pressure, microfluidics stations create an environment in which molecules in a biological sample can essentially "sort themselves out" (Figure 8).

Finally, as a third area of application, current biotech research commonly makes use
of bioMEMS devices for genome and proteomic sequencing and analysis.[17] New lab
technologies such as microarrays (DNA chips and protein chips) not only have shrunk
the molecular biology lab, but have further integrated it with computer technologies.[18] Handheld DNA chips are regularly used in molecular biology labs, and have
been developed for medical use in genetic screening as well. In such cases, the goal is
to further integrate the "wet" data of DNA or proteins with the "dry" data of biological sequence in computer databases. Microarrays such as the DNA chip are—as devices in themselves—hybridizations of biomolecules and technological substrates,

DNA and silicon. But microarrays are also more than that, for they function by transforming DNA into an autodiagnostic molecule, by harnessing the "natural" process of complementary binding between DNA for novel, diagnostic ends (e.g., gene expression profiling). In these instances, bioMEMS highlight the space separating biological data and computer data, or biological and informatic networks. Analytic tools such as microarrays have as their main function the transmission of data types across media, and such devices act as a kind of fulcrum transmitting data from one "platform" to another (Figure 9).

Therefore, a formal analysis of bioMEMS—biosensors, microfluidics, and microarrays—can proceed by identifying both the components (and their properties) and the relationships (and their functions) (Table 2). From Table 2 we can see that there are two basic ways of interpreting microsystems from a philosophical-technical standpoint. One is by prioritizing materials. By focusing on the kinds of materials combined in microsystems, one analyzes the static assembly and construction of unique devices to be used in biotech and biomedicine. This perspective, by focusing on static materials, tends to focus on the combination of organic and nonorganic elements. Culturally speaking, this leads to the trope of the "cyborg," or human–machine hybrid, except on the microscale. Microsystems become noteworthy because they combine biological and technological components in some fashion. Another approach is to prioritize relations between materials. By focusing not just on the components, but the relationships between them, and the processes they undergo, one analyzes not static but dynamic structures. By focusing on materials in action, or rather materials designed as processes (where form is not separate from function), the focus is on the biology–technology mixtures as systems. Culturally, we can say that this leads beyond the cyborg, and toward thinking about "biomedia."

From a materials standpoint, the DNA chip appears to be exactly that—a literal hybrid between biology and technology, DNA and silicon. But this perspective says nothing about how "bio" and "tech" are mixed together; neither does it say anything about what roles each plays. Does technology simply act upon passive DNA? Or, does DNA trigger the technology into activation? From a systems standpoint, however, where the DNA chip is not static but a dynamic system of various processes (silicon-based adhesion, hybridization, digital scanning, software analysis), it becomes apparent that the DNA chip is not simply a hybrid of biology and technology, but the technical contextualizing of biological processes to perform extrabiological tasks.

For microfluidics, the materials standpoint gives us a pure technology, and not a hybrid at all. Microfluidics stations usually do not incorporate biological materials into their design; different micromanufacturing processes etch channels into chips for biological assays. But, as a system, such devices do not operate until they have something that "resists" them, or some substance to interact with. DNA in a microfluidics chip by the company Caliper, for instance, is much more than biology being worked upon by technology; it is a technology that is triggered by biological resistance (mechanical

Figure 9. Affymetrix GeneChip® probe array. Image courtesy of Affymetrix.

pressure, biomolecular properties, polarization). Biology is recontextualized in ways that enable biomolecules to undergo an autosorting, or a biologically driven mode of self-organization (one that is, nevertheless, underwritten by a technical logic).

For bioMEMS, the materials standpoint again gives us something much like traditional medical technology. A technological device measures the body's performance, without interfering in (as in surgery) or replacing (as in artificial organs) that performance. In vivo blood-pressure sensors interface in some way with biological environment, only this time they use IC-based chips to physically detect pressure changes, which are then communicated to the IC chip. From a systems perspective, we can see that the requirement for such devices—especially as long-term implantable devices—is that the right fit or physiological specification must be established between biological and technical systems. The bioMEMS device must integrate itself invisibly into a biological milieu while also providing an interface to that milieu for data generation. This means that the body becomes recontextualized as a kind of pressure machine, which then triggers the bioMEMS device to react.

BioMEMS versus Cyborgs?

What each of these areas of application and their corresponding relationships demonstrate is that bioMEMS devices further integrate biology and technology in novel ways. Despite this, bioMEMS do not represent a literal fusion of biology and technology, nor do they imply the effacement of that boundary altogether. This suggests that the hybrid quality of bioMEMS has to do not only with their material construction, but, more important, with their dynamical functioning within and across different systems.

Table 2. Types of microsystems used in biotechnology research

BioMEMS type	Materials	Function	Data Type	Application
Biosensors	Silicon-based chip with sensors and actuators	In vivo; MEMS device implanted into biological system	Mechanical & electrical signals; computer data (all separate)	Medical diagnosis & therapy; biomedical engineering
Microfluidics	Nonorganic substrate & biological sample	In vitro; etched substrate handles biological sample	Properties of bio-molecules (charge, size, bonding strength); conversion to digital format	Sample preparation for molecular biology research
Microarrays	Silicon-based chip with attached biomolecules	In vitro; hybrid chip analyzes biological sample	Nucleotide or amino acid se-quence; conversion to digital format	Genomics & proteomics; genetic screening

From one perspective, bioMEMS would seem to present us with explicit examples of "cyborgs" for the biotechnical age. These actual fusions between, say, cells and computer chips appear to provide proof of concept for the technocultural concepts of hybridity, assemblages, and the posthuman dream of coevolving humans and machines. In considering these microscale hybridizations between biology and technology, these seemingly literalized instances of cyborg configurations, it is also important to note that bioMEMS are devices engineered to function—that is, to operate dynamically in some conjunction with biological processes. In other words, the merit—technical or otherwise—of a device such as a DNA chip is not so much the fusion of biology and technology as it is the way in which DNA's biological processes (such as hybridization of complementary base pairs) is isolated, engineered into other systems, and then repurposed, so that it may function outside of the cell, but still in a biological context.

This means that, despite their static construction as cyborg entities, bioMEMS devices are, more importantly, contextualized transmissions between varying data types: DNA, micropumps and valves, RNA, miniature electrical circuits, amino acids, silicon substrates, polypeptide chains, and diagnostic computer software. From this view of dynamic systems, we can suggest that at the core of bioMEMS technology is the transmission of data types across different media.

For instance, in the example of biotech research, one possible scenario involves the following: A biological sample (e.g., blood) is taken from a patient. Microfluidics stations employ chip technology to isolate, purify, and amplify targeted DNA that is to be analyzed. The sample DNA, once isolated, is then passed through a microarray (or DNA chip), where fragments of known DNA are attached to a silicon substrate. The DNA chip then "analyzes" the sample through fluorescent-tagged hybridization. The resultant pattern of hybridization can then be scanned by the microarray computer, which digitizes the hybridization pattern (represented as a grid of colored dots), where it can be ported to software for microarray assays. That initial digital pattern is then "decoded" or sequenced according to the known DNA on the DNA chip, and can then be compared to on-line genome databases (such as the human genome projects), to identify gene expression patterns associated with a given genetically based disease.[19]

In this elaborate and routine process, not only are several types of materialities at work, but also several data types are being transmitted, translated, and passed through various media. Such a transmission of data types across variable media is what Lev Manovich, discussing new media generally, refers to as "transcoding." In his analyses of computer-based new media, Manovich suggests that transcoding involves not only the technical file conversion from one data format to another (from VHS to DV; from DNA to ASCII); it also involves the transmission of the metaphors, concepts, and categories of thought from one medium to another.[20]

In this sense, Manovich suggests, data transmitted across varying media not only pertains to what he calls the "computer layer," but it is indissociable from a "cultural layer," or the ways in which media ontologies affect our cultural views. In the context of biotech and bioMEMS research, this multilayered transit of data types—wet data, dry data—moves across many systems and data formats, from DNA's famous four-base system, to the set of twenty amino acids composing proteins, to the digital format in the gene sequencing computer. It is this transcoding between systems, in the interstices of systems, as it were, that is the significant point to highlight with regard to bioMEMS. Rather than focusing on the properties and static construction of bioMEMS, an analysis of the transcoding protocols within bioMEMS can provide us with relationships not just inside a device, but inside the particular system a device enables (e.g., the particular circulatory and physiological system an in vivo blood-pressure sensor enables). This particular relationship then also constitutes a cultural as well as a technical layer, a set of assumptions concerning the signification of DNA (wet DNA, dry DNA) that functionally crosses various platforms.

In one sense, bioMEMS operate through principles of transcoding, as implied in their name—the engineered combination of electrical, mechanical, and biological systems. In this they fulfill the requirements of new media as outlined by Manovich. However, unlike Manovich's characterizations of new media, which are based on a common logic of digitization, bioMEMS face the challenge of potentially incommensurable, resistant, or distorting transmissions across widely varying media. For instance, in the data transmissions between DNA, integrated circuits, and micro-actuators, the issues of histocompatibility and immune-system rejection must be taken into account, and are, from the engineering standpoint, often the most common obstacles to bioMEMS design.[21]

BioMEMS technologies thus present us with a paradox: on the one hand, they cease-lessly hybridize biological and technological components and processes, as in the example of the DNA chip; on the other hand, they maintain the boundary between biology and technology, as implied in their design and application (that is, as "tools" for genomics research). In approaching the relationship between biology and technology, bioMEMS appear to at once hold separate and yet mix together these two domains. How do bioMEMS maintain these two seemingly contradictory positions?

Understanding bioMEMS—and biotechnologies generally—as transcoding protocols can provide us with ways of addressing changing views of the body in biotech and biomedicine. However, this emphasis on transcoding does not mean that the distinction between biological and technological systems drops away altogether. In fact, it is the very difference between systems that is the condition for the transcoding across systems; from an engineering perspective, it is the very peculiarities to electromechanical systems that enable them to transmit electrical signals along a channel to direct DNA fragments onto an array for sample analysis.

Specifically, the key moment in the functioning of bioMEMS has to do with the isolation and repurposing of biological components and processes. This can be, as we have seen, DNA hybridization in genome sequencing bioMEMS or interactions with the flow of cells inside the bloodstream for in vivo probes. In this sense, bioMEMS must maintain a dual functionality, and they do this through a two-step process. The first is the bracketing of biological components and processes (e.g., DNA hybridization in the DNA chip), and the second is the technical repurposing of those components/ processes toward novel ends (e.g., for use in laboratory genome analysis). Biology functions as a technical means, which is nevertheless not separate from "natural" biological function. In such instances, technology is configured not so much as a tool, but as a set of conditions in which biology can act technically upon itself (the silicon in the DNA chip is passive, for DNA does all the work).

DNA and Light

We can illustrate these principles in more detail by examining one class of bioMEMS, the "oligonucleotide microarray" or DNA chip. The DNA chip is most often constructed of a silicon or glass substrate, upon which single-stranded DNA molecules whose sequence is known are attached. A sample of DNA fragments (containing some type of "marker" for identification) whose sequence is not known is then washed over the chip. When DNA binding occurs, this can be detected by identifying the sites on the chip that contain the marker gene (most often a fluorescent tag). Currently, such microarrays are commonly used as part of the molecular biology lab, and their efficiency, high degree of customization, and miniaturization have made possible their use in medical diagnosis, regenerative medicine research, pathogen detection (in biohazard environments), as well as in genomics generally.[22]

To understand the emergence of this technology, it is important to situate the DNA chip in relation to similar techniques in molecular biology. Prior to the integration of computers into the biology lab, gene sequencing was often carried out using the "Sanger method," a technique developed by Fred Sanger in the 1970s.[23] The Sanger method makes use of two separate techniques, both of which are still used in biotech research employing gene sequencing today.

The first technique is that of "restriction mapping," which is the use of specialized "cutting" enzymes that snip DNA at specific points. Such enzymes are known as "restriction endonucleases" or simply restriction enzymes. They are specialized bio-molecules that will cut DNA when they recognize a given sequence or pattern. The restriction enzyme *EcoRI*, for instance, which is often used in E. coli studies, will cut DNA whenever it recognizes a certain sequence (such as ATTACG). Researchers use restriction enzymes to locate gene regions of various fragment lengths, which can be a useful tool for identifying genes involved in disease, for instance. A version of this basic technique involves an additive, rather than subtractive process. This is known as "chain-termination sequencing." Whereas the use of restriction enzymes aims to cut an already-assembled DNA chain, the chain-termination technique aims to construct DNA chains of variable length. The synthesis is facilitated by an enzyme known as "DNA polymerase," which simply does the opposite of the restriction enzyme. When complementary base pairs are present, the polymerase will "glue" them together. This process of DNA synthesis is stopped or terminated each time the DNA chain being assembled runs across an "incomplete" nucleotide, known as "dideoxynucleotides" ("deoxy" because they are missing oxygen molecules, "di" because there are two sites missing oxygen molecules). These are intentionally placed in the medium, and are lacking an oxygen molecule, and this difference makes it impossible for the DNA synthesis to occur (it is an intentionally designed "broken" nucleotide). Because of this, researchers know that adding ddATP, for instance, will terminate DNA synthesis every time the assembled DNA chain reaches a "T" or thymine. By collecting enough of these fragments (e.g., DNA chains that stop at every "T"), researchers can method-ically derive the sequence of the sample DNA being studied (Figure 10).

The second technique is that of "gel electrophoresis," which is the use of specialized gel substrates (most often a sugar-based composition such as agarose) that selectively allow the passage of biomolecules according to their size. The gels, viewed microscop-ically, are actually dense matrices, which function like a filter or sieve. This gel, a layer placed between two glass slabs, can be charged, by placing electrodes at either end. Because DNA is a highly negatively charged molecule (owing to the presence of phos-phate groups), it will move toward the positive charge, through the gel. Large DNA molecules will, therefore, have a tougher time moving through the gel than shorter ones will, resembling a kind of molecular DNA race, with different samples in differ-ent lanes. Once a DNA sample is placed in the "wells" (deposits at the end of the gel), and the gel is "run" (electrical current is applied), the gel itself will contain a snapshot of the process. In order to get the results, the gel itself must be stained, so that the individual locations of the DNA fragments can be seen. The use of solutions such as ethidium bromide (EtBr) is common, because it binds to DNA molecules, and emits a

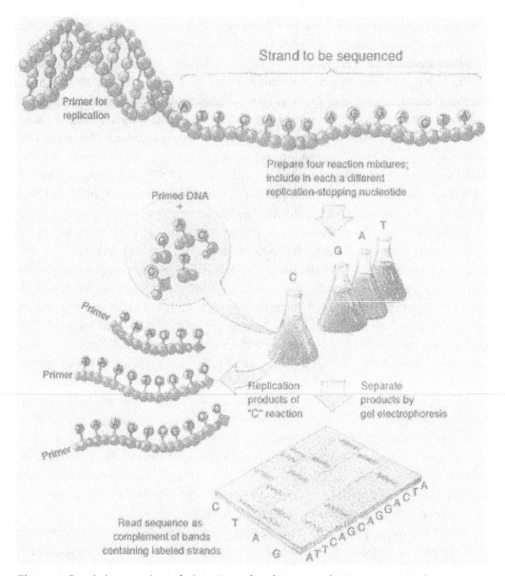

Figure 10. Restriction mapping technique. Reproduced courtesy of U.S. Department of Energy Human Genome Program, http://www.ornl.gov/hgmis.

fluorescent glow when placed under ultraviolet (UV) light. Once this is done, the locations of DNA will appear as glowing bands. The bands closer to the negatively charged end of the gel (the "start") will therefore represent larger DNA molecules, while the ones closer to the positively charged end (the "finish") will represent the smaller molecules. If nothing about a sample DNA is known except its general size (number of bases), gel electrophoresis can be used to assess whether or not a sample DNA contains the sought-for gene. If a partial knowledge of the sought-for sequence is known, electrophoresis can likewise be used to help predict where along the sample DNA that gene is located.

The Sanger method brings together the techniques of chain-termination sequencing and gel electrophoresis, into a multistep process that enables the sequence of a sample DNA to be elucidated. It is important to note that the key to the Sanger method (as well as its two main techniques) is the knowledge about how DNA nucleotides (adenine, guanine, cytosine, thymine) bind to themselves. Known generally as "base pair complementarity" or "Watson–Crick complementarity," this simply states that, under normal conditions, A will always bind to T, just as C will always bind to G. This means that the basic knowledge of DNA base pair complementarity (A binds to T; C binds to G) provides the foundation for a whole set of techniques for working with DNA; that is, the "biological" principles of base pair complementarity, which commonly take place in the cell, provide the occasion for their "technical" redeployment (in gene sequencing).

The Sanger method is composed of four basic steps. The first step is the preparation of the sample DNA. This DNA is denatured so as to form a single strand, and is replicated many times. Each sample DNA is labeled by a "primer," a short sequence that is radioactively tagged, for easy identification. Along with the sample DNA are added free nucleotides ("deoxynucleotides," where an oxygen is removed to make them "sticky") in equal quantity. Four separate solutions are made, each corresponding to one of the four nucleotides. In each solution, a small amount of dideoxynucleotides (the "broken" nucleotides) is added: to one solution, the dideoxynucleotide corresponding to thymine (ddTTP), to the second, the dideoxynucleotide corresponding to cytosine (ddCTP), and so forth. Therefore, in each solution, a specific set of terminating molecules is active, with everything else being equal.

The second step involves heating and cooling procedures to enable DNA synthesis in each of the four solutions to occur, by adding the "glue" of DNA polymerase. Again, the polymerase will continue assembling DNA chains until it reaches a dideoxy- or broken nucleotide, at which point DNA synthesis will stop. After this, the third step involves porting the chain-termination process into gel electrophoresis. The solutions of DNA (now synthesized, but constrained by the dideoxy- elements) are then run through gel electrophoresis as described earlier, with the longer DNA molecules moving slower and the shorter ones faster. The final step involves reading the results, by staining the gel and aligning the results from each solution alongside each other. Recall that, because the chain-termination process results in a series of terminated DNA fragments (e.g., the ddATP solution will have fragments that are terminated every "A," etc.), there will be a fragment for every position in the sample DNA covered by one of the four solutions. Knowing this, researchers can simply read the gel from top to bottom, scanning each column (the "A" column, "T" column, etc.). The resultant sequence will therefore be complementary to the unknown sample DNA sequence. By simply reversing the resultant sequence, you have the sequence of your sample.

This process is extremely time-consuming, and prohibits the running of simultaneous assays. One approach has simply been to automate the Sanger method (this is, in fact, the basic technique used in current gene sequencing computers); another, similar approach has made use of fabrication techniques used in the manufacturing of integrated circuits for computers. This approach has resulted in the microarrays or DNA chips marketed by companies such as Affymetrix, Agilent, and Nanogen. In a 1993 research article in *Nature*, "Multiplexed Biochemical Assays with Biological Chips," an Affymetrix research team headed by Stephen Fodor suggested that the manufacturing techniques used to etch circuits in silicon chips could also be used to manufacture miniature "biological chips" for DNA analysis.[24] A later article built on this knowledge to specify how these DNA chips might be used in molecular biology research, placing particular emphasis on the construction of customized probe arrays.[25] In these articles, Fodor et al. specifically point to two technologies that would make DNA chips possible: "light-directed combinatorial chemical synthesis" and "laser confocal fluorescence detection." The former derives from research in physical chemistry, and is the use of photolithographic techniques to chemically bond specified molecules to a glass or silicon surface. The latter is the use of fluorescence tagging of biomolecules for post-assay identification, and has been a mainstay of much biotechnology research for some time, though in the case of DNA chips this identification is aided by digital computers. In between these two technologies is the actual assay itself, which involves the use of base pair complementarity (used in the Sanger method) to assess whether an unknown sample is a "match." Therefore, in early articles on DNA chips, Fodor et al. already identify a three-step process: chip assembly (using IC-based photolithography), biological assay (using base pair complementarity), and output of results (using computerized fluorescence tagging).

In the first step of chip assembly, the sample principles used to manufacture computer chips are redeployed for DNA chips. The primary difference, however, is that whereas IC chip photolithography is primarily a "negative" process, etching away a surface to reveal circuits, the process in the case of the DNA chip is both negative and positive, because the etching only occurs to enable the attaching of DNA molecules. Using a photolithographic "mask," light is directed onto the chip surface, eating away the surface layer ("photodeprotection") and making it available for chemical bonding. To these deprotected sites on the surface, any range of chemically prepared biomolecules may be attached, including DNA and proteins. This basic process can be repeated any number of times, and each application can involve a different mask, as well as the attachment of different molecules to the surface. By selectively deprotecting and chemically coupling molecules to the chip surface or strands, chip-attached oligonucleotides can be literally assembled base by base. This use of combinatorial chemical bonding enables a customizable array by forming a grid onto which oligonucleotides are "spotted."

Because the sequence of the oligonucleotides that are spotted on the grid is known (e.g., cell #4 may contain a single-strand DNA of ATTACGCAG), they function as "probes" or points of reference for DNA samples whose sequence is unknown. This second step, which is the actual biological assay, takes advantage of base pair complementarity by washing a sample over the DNA chip, or by immersing the DNA chip in a solution in which sample DNA fragments are included. Because the DNA strands attached to the chip are single-stranded (or "sticky"), any complementary strands will bind to them somewhere along the DNA strand, forming complete or partial hybridized DNA. The detection of the assay results is enabled by the third step, which uses a form of light detection (laser confocal fluorescence detection) that is able to identify the matches on the chip by illuminating those spots where DNA has double-bonded, revealing the radioactive tag as a particular color. An important part of this step is the use of computer technologies (automated fluorescence scanning and digital imaging) to digitize the microarray, and produce output images of brightly colored, spotted grids.

In the initial experiments by Fodor's Affymetrix team, a test sequence of DNA composed of twelve base pairs (AGCCTAGCTGAA) was attached to a chip. A sample of unknown DNA fragments of eight base pairs of different sequences was then washed over the chip. Each type of sequence is, recall, radioactively labeled. After fluorescence detection, the team was able to detect which of the cells on the chip grid contained a match. Because the sample DNA may bind to the test DNA at any one of its locations, each match will not be a perfect match, but will represent a partial hybridization, beginning at one of the test sequence's initial points. Because sets of the unknown DNA contained eight base pairs, and the test (known) sequence contained twelve base pairs, the Affymetrix team estimated that five of the unknown sequences would most strongly bond to the test DNA. At each cell in the grid where a spot occurs, a match (DNA hybridization through base pair complementarity) has taken place. Knowing the test sequence at each cell also implies the sequence of the unknown sample that has bound to it. Gathering all of these sequences together, the Affymetrix team could then align them, to reconstruct a total complementary sequence. This sequence (in this case, a complementary to the twelve-base test DNA) can then be easily converted, giving the test sequence with which the experiment began. This process has been referred to by Fodor et al. as "sequence by hybridization" (SBH).

With this specific technique in mind, the DNA chip would seem, at first glance, to be a literal flesh-and-silicon cyborg, a molecular version of the laboring body (where the molecular factory is the sequencing lab). This is, in part, true, but we need to reiterate that the DNA chip, in its structure and function, actually combines two types of DNA—that of the biological sample to be analyzed, and that on the DNA chip itself. Between them, microelectronics technology and bioinformatics software are utilized as the means of setting up particular (biochemical) conditions, so that DNA rubs up against DNA; that is, biotechnologies such as the DNA chip are less about the instrumental uses of computer technology, and more about the ways in which DNA technically works upon DNA. There certainly is a unique hybrid of DNA and electronics going on here, but equally important are the ways in which the techniques of the DNA chip demonstrate the technical uses of biological elements and "natural" cellular processes. Once a particular set of conditions is set up, designed, for particular uses—

in this case, for the hybridization of single-stranded DNA segments—we are no longer talking about machines working on bodies, but rather bodies interacting with each other under certain laboratory contexts.

When we look more closely at the techniques of biotech, we see not just an instrumental approach to technology, and not just the production of cyborgs, but a complex investiture in the capacities for technically recontextualizing the biological body. By working on the level of informatics, where genetic and computer codes evince a reciprocal relationship of transcoding, biotech is actually transforming biology itself into a technology. When we consider the DNA chip on the level of its material practices (sets of procedures that are articulated in a discourse toward the fulfillment of certain goals), we can see that it is a technically contextualized site in which DNA works upon DNA, at once "natural" (base pair matching) and "artificial" (bioengineered samples and labeling of hybridization patterns).

In our consideration of the DNA chip, we can see something happening that is more than the literal intersection of biology and technology. In a sense, there is very little "technology" (in its common connotation as a "black box" or "hardware") in the DNA chip. The main work in DNA sample analysis is carried out not by any computer or silicon-based system, but by DNA itself, through DNA's "natural" tendency to bind complementary strands of nucleotides (the natural tendency of adenine or "A" to bind with thymine or "T," and so on). This general approach can be found in other examples of bioMEMS as well—the self-organization of microfluidics stations, and the semiautonomy of implantable biosensors. In a sense, then, the DNA chip is significant not because it represents a literalization of flesh and machine, but because "technology" seems to mysteriously disappear altogether, or at least to fade into the background, highlighting the conditions in which the genetic body can perform its "natural" propensity to regenerate, replicate, or differentiate.

The feasibility study of Fodor et al. for the DNA chip aims to show a number of things. The first is that biomolecules such as DNA, RNA, and proteins can be chemically coupled with "nonbiological" substrates without losing their biological functionality (base pair binding). Extending from this, these studies also demonstrate how a biological assay can be "translated" from one medium to another (from the gel electrophoresis to the DNA chip, from the DNA chip to the computer). Although the Sanger method and gel electrophoresis are as "technological" as the DNA chip, the way in which the DNA chip studies enframe the biological assay is important to understand, for it constitutes a "biomedia" that crosses several material substrates: the "traditional" media of the wet lab (gel electrophoresis, fluorescence tagging), the media of IC chip manufacturing (photolithography), the hybrid media of the microarray itself (biological and nonbiological components), and the digital medium of the computer (microarray scanning, imaging, and analysis software). For Fodor's Affymetrix team, what remains constant across these media is the central element of the biological assay, based on the principle of base pair complementarity (or, in the case of chain-termination sequence, the failure of base pair complementarity). This principle, which is activated on the hybrid chip itself, is also worked upon at each stage of the DNA chip technique: the preparation and assembly of the custom probe array creates the infrastructural context in which DNA can perform its "bio-logic," just as the output stage of computerized fluorescence detection and analysis reassembles and reconstructs DNA as pure sequence (e.g., the computer algorithms that align the matches to form a single, total sequence).

"Remember, I need oxygen"

In terms of biomedia, we can see how the DNA chip is literally enframed (by a grid) by selected technologies. The principle of base pair binding, taken as both "biological" and "natural" (regularly occurring in living cells), is converted into a bio-logic through its being enframed by two "light-technologies": photolithography and fluorescence detection. DNA framed by light, molecules framed by bits. In the specific case of the DNA chip, and in the case of bioMEMS generally, we see not so much cyborg intersections of flesh and machine as technically recontextualized bio-logics. The "technological" is, in these cases, not dependent on its ability to produce, to change, or to augment, but on its ability to enhance what always remains biological.

It is this technical contextualizing of "natural" biological processes in biotech research that we have been referring to as "biomedia." *Biomedia* is a term that describes the ways in which biology is reconfigured as a technology; in this it is one of the defining characteristics of biotechnology itself, as the technical conditioning of biological components and processes. For bioMEMS, the approach of biomedia is to combine biological with electromechanical principles, in order to generate certain types of data. This very logic puts forth several assumptions: first, that data inheres in "wet" biomolecules in a way that is commensurate with "dry" computer systems; second, that bioMEMS devices simply make explicit what was previously implicit (the "natural" data of the organism); and third, that the body's data types are extensible (that beyond this natural data, there are other data types that can be technically developed, such as genomic profiling or molecular simulation).

In one sense, bioMEMS are not unique, in that implantable devices, artificial organs, and prosthetics have long been the domain of biomedicine. What is unique about bioMEMS is their manifold intersections with new media (such as computer science and integrated-circuit technologies). BioMEMS are not exactly medical tools (as are the X ray, CT, or MRI), in that part of what makes them function is the integration of living biological components; similarly, bioMEMS are not exactly intended to be biological replacements (as are artificial organs), in that they are engineered for the purposes of analysis and diagnostics. BioMEMS seem to hover somewhere between living and technological systems. BioMEMS are a specific example of biomedia in that they explicitly bring together the body and technology into a relationship where the biological may continue to function as biological (through the interdisciplinarity of bioscience and engineering). They recontextualize the biological body as a technology in three primary ways.

1. Through a device-environment relationship: By integrating themselves into the biological milieu of the body, bioMEMS transform the body's biomolecular operational interior (blood-pressure levels, metabolic activity, drug-delivery monitoring) into a system of optimization. Muscle stimulators, artificial organ MEMS, and optical prostheses all involve minimally invasive devices that simply, through their mechanical and electrical structures, conduct the flow of matter and energy in a localized biological system. BioMEMS introduce a novel form of systemic regulation at different levels of the biological body's functioning (blood sugar levels, poor muscle conductivity, etc.).

2. Through a biological-mechanical relationship: By developing novel hybrids between biological and mechanical components, bioMEMS form a literalized instance of a biomechanics. Especially in related fields such as microarrays, microfluidics, and living bioMEMS, the boundary between organic and inorganic, natural and artificial, and human and machine is brought into an ever more intimate relationship. However, the biological components do not simply function as they did before, partly because they have been removed from a particular milieu, and partly because their bio-"logic" has been extracted and fused with the mechanico-logic of an MEMS device. The resultant bioMEMS assembly may serve a biological function (e.g., in facilitating neural signal transduction), but it will be a unique biological function that is new to both the mechanical MEMS device and the biological components hybridized to that device.

3. Through bioinformatic networks: For bioMEMS applied to either biotechnology research (e.g., in vitro or in vivo gene expression profiling) or biomedical diagnostics (e.g., monitoring the patient's body), signal processing will be a crucial technical issue, ensuring that MEMS sensors can output relevant signals to an external device (most likely another set of sensors outside the body, or to a computer). For research, this means using bioMEMS to efficiently and accurately analyze large amounts of complex biomolecular aggregates (genes, proteins), sending data from bioMEMS devices to computers for data analysis. For biomedicine, this means exploring wireless technologies to enable semi-long-term implantable biosensors that monitor a patient's biological output (blood sugar levels in diabetes patients, for example). In either case, a network is established, both in the digitization of biomolecular interactions (translatability of molecules into data) and in the externalization of informatic signals from the body's interior (McLuhan's "extensions of man").

The examples included in the bioMEMS class—biosensors, microarrays, and microfluidics—are each examples in which the relationship between bodies and technologies is reconfigured in a particular way. In their literalized hybridization between "wet" biological materials and "dry" nonbiological technologies, microsystems provide an example of a specific materialization of the trope of the "cyborg." However, despite the fact that microsystems directly bring together organic and nonorganic components, this intersection is not simply a fusion or a literalized human–machine assemblage. Each type of device fashions cyborgic instances in significantly different ways, sometimes operating according to a device–milieu relationship (biosensors), sometimes forming direct hybridizations (microarrays or DNA chips), and sometimes constituting structures for the "information processing" of biomolecules (microfluidics or labs on a chip).

Metadesign and Metaorganisms

BioMEMS constitute a class of material devices that raise a whole set of philosophical questions pertaining to the body and technology. Do these philosophical questions have any relevance in relation to bioMEMS and biotech? The suggestion being made here is that they do, for they form the ontological basis from which principles of biomedical engineering begin, and from which a broader cultural understanding of the body in biotech and biomedicine may emerge.

As we have seen, bioMEMS are an example of a set of practices in the development of biotechnologies that at once hold separate and also fuse the boundary between living and nonliving, biological and nonbiological. In this they harbor a tension in the status of the device vis-à-vis its context. Examples such as in vivo biosensors and DNA chips suggest that device design should proceed not from the engineering perspective of tool (because part of the tool is composed of living biological components), nor from the biomedical engineering perspective of functional replacement (because the purpose of the device is not biological but biodiagnostic). Rather, we might think about the status of devices from the perspective of integrated systems that may cross several boundaries: integrated circuits fused with living cells that analyze biological samples that are then uploaded into a database.

From this standpoint, it is important to ask, even if speculatively, whether or not these complex intersections between "biological" and "nonbiological" components and processes also implies transformations to what counts as "living" or "nonliving." From one perspective, bioMEMS are simply tools—that is, at least, how they are currently used. From another perspective, bioMEMS as a relationship potentially reformulate how the "biological" is conceived of from a biotechnical, engineering standpoint. The given of the organism or the cell gives way to a living system amenable to modes of technical intervention and regulation; the organism becomes a "metaorganism."

The metaorganism is not a technically enhanced cell or organism, but the process by which the body, inscribed by biotechnologies, is elevated beyond a state of mere biological "animality." Historically, this has been achieved by an increasingly sophisticated mechanistic account of the body (not a messy organism, but a complex, smoothly running clockwork machine), which has been repurposed today in terms of informatics.[26] The metaorganism is the body beyond itself; it is a mode of materiality producing a discourse around the human body, through which the body may be "elevated" technically, as well as conceptually and ethically. This is not the transcendence of the body, nor is it the manifestation of anxiety toward the body. In the case of bioMEMS, the body is only despised insofar as it does not fulfill itself: base pair complementarity is only a problem when it does not function within the infrastructure of the microarray. By "optimizing" the body, what first must happen is a qualitative maximization in the modes whereby the body becomes an object of knowledge, or better, a resource of information.

By extending the modes of informatic knowledges of the body, fields such as bio-MEMS proliferate and add dimensions to the bodies of biotechnology. However, this is not simply a decentralizing of a single, hegemonic (anatomical-medical) discourse. In many ways, the broad task of biotech research is, on the one hand, to proliferate the points of contact between informatics knowledge and the body and, on the other hand, to channel these different, sometimes heterogeneous knowledges back to a model articulated by preexisting, dominant modes of introducing new technologies into medical practice, diagnostics, and research. The more points of contact, the more the discourse of biotech can become infused with the "natural" body, the more it can converge toward an ideal point where the two overlap, and where naturalization and transparency occur to such an extent that the biotechnical is in fact the natural.[27] This increase in the points of contact with the body implies not only a proliferation and diversification of knowledge about the body, but also a feedback system in which knowledge produces the conditions for the very possibility of the body's official, legitimate, instantiation.

The concept of the metaorganism, or the organism-more-than-animal, is meant to suggest that, historically and in contemporary instances, fields of biotech practice unique modes of approaching, knowing, and recontextualizing the body. Again, this is not so much a transcendentalist science of the body, in that there is not a will to devalue, dispose of, or repudiate the body. Although, in numerous examples (the Human Genome Project being the most noteworthy), some transcendental principle does form the grounding for the study of the body, the primary, operational telos of biotech is not a principle of mind, soul, or consciousness, but rather the understanding of the body as a passageway to a range of other concerns: as a way of understanding the inherent, secret order of the body; as a means of optimizing the biological body; as a means of a more complete regulation of the genetic body; as a means of seeking the genetic basis of disease. In formulating the metaorganism, biotech is primarily interested in the generative capacities of the body, as it is invested in the informatic order of the body, compelled to understand an object that is at once part of the subject and distinct from it (being, becoming, having a body). Many processes are involved here, none of which holds claim to being the whole of biotechnology: abstracting, encoding, recoding, decoding, "anatomizing," universalizing, fragmenting, organizing, assembling, and so forth.

The principle of the metaorganism is therefore the elevation of the biological through technologies of knowledge production. Through this principle, contemporary biotech in general, and bioMEMS specifically, have two main effects on the notion of the biomolecular body. One effect is that contemporary biotech research begins from a technical position in which the body (the "natural" body, in which the biological is implicit) is thoroughly valued as a source of knowledge. Despite its in-depth engagement with computer technologies, biotechnology remains a practice invested in the body as nature, and as biological materiality. The second effect is that, through its integration of unique technologies, biotech increases the points of contact on the body, thereby increasing the dimensions of knowledge possible on the biomolecular body. New techniques make for new, increased, and diversified knowledges of the body, which in turn make for novel modes of optimizing "bio-logics." It may be said that these knowledges, and the bodies they articulate, are, in an important way, in excess of the embodiedness of patients, biological samples, and biomedical subjects, in the same way that virtual bodies or computer-generated bodies exceed the ontological boundaries of individual subjects in the "real world."

This is to suggest that contemporary biotech "optimizes" the biomolecular body, and that this optimization involves the use of the knowledges gained from new technologies toward the increased accuracy, diversity, efficiency, and expandability of the human body as a biomolecular entity. New diagnostic techniques, new computational modes of analysis, new modes of visualization, new methods of organizing data, all contribute to a tighter, more complex, and more sophisticated knowledge of the body. Contemporary biotech proliferates sets of norms as a result of this optimization. These norms are not monolithic, nor are they deterministic. Their flexibility lies in the capacity of the technology to instantiate universal models (e.g., the "bio-logic" that says DNA is DNA, whether in the body, the lab, or part of a bioMEMS), as well as the modeling of individual patients (e.g., customized DNA chips for patient-specific genotypic analysis).

What is instructive about the examples of bioMEMS is not that they realize the science-fictional dream of fusing human and machine, nor that they enable more seamless modes of crossing boundaries (DNA-silicon, molecules–light). BioMEMS are noteworthy because, in their design and implementation, they inadvertently raise the question of how the "biological" is defined in the biotech era of DNA chips, bioinformatics, and nano-molecular circuitry. On the one hand, bioMEMS-related devices are clearly designed as analytic and diagnostic tools for biomedicine and biotech research (implied in their projected uses and functions); they are different from biomedical engineering projects that aim to replace, improve, or redesign the biological body, and in this sense they are "body tools." On the other hand, bioMEMS themselves are clearly novel configurations of the relationship between the biological and technological domains, as they are traditionally tied to the divide between organic and nonorganic materials. The level of success for bioMEMS (be it in vivo sensors or handheld microarrays) will be the ability of the devices to function as biological systems (e.g., DNA hybridization, in vivo integration), in spite of, or because of, their hybrid design. BioMEMS are an example of a set of practices in the development of biotechnologies that at once hold separate and integrate living and nonliving, biological and nonbiological. In this they harbor a tension in the status of the device vis-à-vis its context.

Such boundary regulations raise questions that are simultaneously philosophical, technical, and, in the end, bioethical: Is a long-term in vivo biosensor part of the patient's biophysiology? How does a patient's own DNA affect how that DNA samples itself in microarrays? Examples such as in vivo biosensors and DNA chips suggest that device design should proceed not exclusively from the engineering perspective of tools (because part of the tool is composed of living biological components), nor exclusively from the biomedical engineering perspective of functional replacement (because the purpose of the device is not biological but biodiagnostic). Rather, we might think about the status of bioMEMS devices from the perspective of integrated systems that may cross several boundaries: ICs fused with living cells that analyze biological samples that are then uploaded into a database. From a theoretical perspective, bioMEMS are illustrative of the need for biotech research to maintain an emphasis on the embodied, situated quality of biotechnical hybrid systems. Both from the perspective of research and design, and from the perspective of critical contextualizing of biotech, there is a need for a means of working beyond the irresolvable biology–technology boundary, a need materialized in the very workings of bioMEMS themselves.

CHAPTER FOUR

Biocomputing

Is the Genome a Computer?

The Life of a Computer

What if "life" turned out to be a form of computation? "Wang's Carpets," a short story by Greg Egan, explores the question of how the boundary between biology and computers may be negotiated in the future.[1] Revised as a section of the novel *Diaspora*, "Wang's Carpets" replays a familiar trope in science fiction: the search for alien life, which is also a search for the criteria for alien life. Paolo Venetti, a sentient software "citizen," comes across a unique type of life-form on the planet Orpheus, some twenty light-years from Earth. The life-forms appear to be immensely large, planar, kelplike "carpets" covering the planet. The description that Paolo's computer gives of the life-form suggests that it is far from being living, let alone sentient:

> The carpet was not a colony of single-celled creatures. Nor was it a multicellular organism. It was a single molecule, a two-dimensional polymer weighing twenty-five thousand tons. A giant sheet of folded polysaccharide, a complex mesh of interlinked pentose and hexose sugars hung with alkyl and amide side chains.[2]

Upon further analysis, Paolo discovers that the carpets—called Wang's Carpets—display both computational and biological characteristics. Not only do they function like autocatalytic enzymes, ceaselessly generating and regulating themselves, but their planar structure also makes them a kind of Turing machine, able to perform simple calculations based on the combinations of the different tiles in the carpets. As Hermann Karpal, another character in the novel, notes to Paolo, "they [the Wang's Carpets] can calculate anything at all—depending on the data they start with. Every daughter fragment is like a program being fed to a chemical computer. Growth executes the program."[3]

However, the Wang's Carpets do not simply carry out their "computations" arbitrarily. Paolo and Karpal discover that their view of the large, planar structures is but a fraction of a polydimensional space encoded by the Wang's Carpets, in which organisms they refer to as "squids" interact and communicate with each other through physical contact. Eventually, Paolo and the others with him must confront a question both highly abstract and immediately political. Not only do the Wang's Carpets demonstrate both computational and complex biological characteristics, but the recursive way in which they do this raises the issue of sentience:

> "All the creatures here gather information about each other by contact alone—which is actually quite a rich means of exchanging data, with so many dimensions. What you're seeing is communication by touch."
> "Communication about what?"
> "Just gossip, I expect. Social relationships."
> Paolo stared at the writhing mass of tentacles.
> "You think they're conscious?"
> Karpal, point-like, grinned broadly. "They have a central control structure, with more connectivity than a citizen's brain, which correlates data gathered from the skin . . . Right or wrong, it certainly tries to know what the others are thinking about. And"—he pointed out another set of links, leading to another, less crude, miniature squid mind—"it thinks about its own thoughts as well. I'd call that consciousness, wouldn't you?"[4]

Aside from the difficulties of picturing polydimensional organisms encoded by kelplike structures the size of continents, the very concept of the Wang's Carpets takes on a different tone in the ensuing chapters. Both Paolo and Karpal must decide how to present this discovery to the "polis," or governmental body, which will make decisions concerning whether or not to further explore and possibly colonize other planets. At the center of this issue is therefore the question of designating the Wang's Carpets as living or nonliving, as biological or computational, or as something both and neither.

The very idea of creating a computer out of biology seems at once an invention of science fiction and a curiously old idea. As Egan's short story illustrates in detail, the question of whether the "living" excludes the "computational" tends to become vague in the discourses of connectionist cognitive science, artificial intelligence, and related fields in which the brain is technically related to microelectronic computing machinery. However, histories of the modern computer point to the fact that, during the era of business industrialism, a "computer" was indeed a person, most often a person sitting in front of a mechanical tabulator, manually performing statistical calculations for the census or for a company's books of production and sales.[5] We can even extend this nominalism further and suggest that the very idea of technology, especially when aligned with labor, implies a reconceptualization of the living human body as a machine—we work to think/calculate and think/calculate to work. The notion of the "biocomputer" is therefore a case of something forever arriving in the future, as well as something that has always existed in principle. It elicits a range of heterogeneous images, from a new type of "bioport" technology to the image of outsourcing our brain-ROM for hire.[6]

However, within the discourses of discrete mathematics and computer science, "biocomputing" is becoming an increasingly specific set of both theoretical and practical procedures, with specific aims and questions raised within its cross-disciplinary context. One way of beginning to discuss biocomputing (or biological computing) is to differentiate it from a seemingly related field, that of computational biology or bioinformatics.[7] On the surface, there would seem to be little difference between biological computing and computational biology. There are two important differences, however. The first has to do with the chosen medium of each. In biological computing (or biocomputing), the chosen medium is biological, with particular emphasis on the properties of DNA and protein molecules and interaction processes. In computational biology (or bioinformatics), the medium is the computer, with particular emphasis on the use of computer technologies to simulate, model, and analyze data that is biological in its origin.

If that were the only difference, then both biological computing and computational biology would simply be two approaches to doing biology, akin to the difference between traditional "wet lab" techniques (e.g., cell culturing) and current "dry lab" approaches (e.g., computational modeling). However, a second difference can be added to the first, and that is the difference in conceptual aims. If computational biology (bioinformatics) has as its aim the use of computers to extend biology and biomedical research, biological computing (biocomputing) has as its aim the use of biology to extend computer science research. In short, we can say that computational biology/ bioinformatics makes a biological use of computers, while biological computing/ biocomputing makes a computational use of biology. For the sake of clarity, we will from here on out refer to biological computing as "biocomputing," in order to avoid terminological confusion. We will refer to biocomputing as the use of biological components and processes toward nonbiological, computational ends.

From this basic distinction (a distinction of disciplines as well), we already see two central characteristics of biocomputing: its overall conceptual aim of developing novel computational technologies, and its means of doing so through the use of biological components and processes. It will be helpful to contextualize these characteristics by reference to two rather well-known developments in computer science that have served as the main motivations of biocomputing. These are computer storage capacity and parallel processing. The former refers to the limits often placed on current silicon-based computers in terms of their storage capacity. The increasing miniaturization of computer storage technologies has led some to speculate on the possibilities of building nano-scale storage devices. However, it has been known for some time that one of greatest data storage devices exists in each of our cells—the intricately compacted, densely coiled, and elaborately indexed molecule of DNA. From a computer science perspective, the ability of DNA to act as an ultraspecific, dynamic database, as well as its large storage capacity, makes it an ideal model for studying computer storage in biological media. This interest is bolstered by another development, which is in computer processing technologies. The ever-present shadow of Moore's Law, which refers to the progressive shrinking—and limit—of integrated-circuit (IC) technology, has prompted many to speculate as to whether other media might serve as inheritors of the microelectronics in current computers.[8] Researchers in biocomputing have proposed that DNA's base-pair complementarity (A-T; C-G) offers not one but two binary pairs, which could be used together for massively parallel processing applications.[9] In fact, the first proof-of-concept experiments in biocomputing were applied to types of computational problems that have, in the past, provide difficult, if not impossible, for traditional silicon-based computers.[10]

Therefore, the impetus for biocomputing emerges not from biology or biotechnology, but rather from the computer industry. The basic goal of biocomputing would then be to make use of biological components and processes (e.g., DNA base pair binding, protein folding, cell signaling) in an explicitly "nonbiological" manner. What would be the applications of such biocomputers? For one, the majority of researchers and spokespeople for biocomputing are clear in that biocomputers will most likely not replace our familiar desktop computers. Most foresee the possibility of either hybrid systems (with two types of processors, a silicon one for linear computations, a biological one for parallel computations) or fully biological computing systems geared toward highly specific problem application. The candidates tested thus far include cryptography and mathematical problems for which an exponentially large search field exists (we will turn to these examples later).[11] As may be expected, both the military and the IT industries have expressed a limited, cautious interest in biocomputing, though mostly at the level of R&D.[12] The majority of researchers working in the field are thus academic researchers, predominantly with backgrounds in either computer science or discrete mathematics. Rare are the biocomputing initiatives with biologists or within biology or biomedical departments.

Again, the notion of biocomputing seems vague—exactly how does one make a computer out of biology? Before addressing the details of biocomputing, it is important to assert that biocomputing is first and foremost an approach to the idea of "computability." In this, biocomputing is as much a conceptual practice as it is a technical one; or rather, we will want to stress the intertwined nature of the conceptual and the pragmatic in biocomputing research. In this way, we can continue by pointing out one of the central principles of biocomputing: that there is an equivalency between the notion of the "biological" and the "computational" that is based on a dissociation of medium and process. In other words, the motto of biocomputing may be paraphrased as follows: the genome is a computer. By this we mean that, for biocomputing, there are characteristics of computing that are seen to inhere in biological components (such as DNA) and processes (such as protein binding to cell membranes). This raises one of the fundamental—and still debated—issues in the very idea of biocomputing: how "computability" is defined, and, by implication, how "biological" is defined; for, as we will see, biocomputing puts forth a set of material and pragmatic claims concerning the ways in which the biological and computational domains can be reconfigured and redesigned.

Biocomputing, as of this writing, is generally divided along three lines, according to the type of biological components and processes utilized.[13] The first biocomputing experiments were performed using simple DNA fragments and their base pair binding characteristics. This area of DNA computing is, as we will see, based on the parallel processing capacities of DNA's double binary set. Researchers encode elements of a problem to be solved into the DNA (for instance, a problem concerning the best route between multiple destination points), and, using standard molecular biology laboratory techniques, allow the DNA fragments to selectively bind to each other. From this biologically enabled combinatorics, DNA base pair binding "selects" and solves the problem.[14]

A second area, alongside DNA computing, is that of membrane computing. While DNA computing is based on the linearized matching of strings (DNA strands), membrane computing makes use of the ultraprecise molecular "fit" between specific protein transit molecules and protein membrane molecules. This lock-and-key structure ensures the passage of molecules into the cell (such as nutrient molecules) and prepares the delivery of molecules out of the cell (such as enzymes produced inside the cell). The cell membrane therefore has a kind of cascading mechanics to it, in which molecular fit between receptor regions on a protein and a membrane initiate a series of chain-reaction events that result in the membrane's changing shape and the entry or exit of the triggering molecule.[15]

A third and more recent area is that of cellular computing, a broadly defined area that includes processes such as cell signaling (akin to the process in membrane computing), protein-protein interactions (folding and binding between protein structures), and cell metabolism (the particular "pathways" of a given chemical reaction). As the most theoretically oriented of the biocomputing fields, cellular computing is unique in that it places emphasis less on components (DNA, proteins, membranes, cells), and more on the inherent network properties of cellular processes. If enough is known about a given biochemical reaction in the cell, then, theoretically, that reaction pathway can be used as a kind of distributed processing network within (and between) cells[16]—if, however, enough is known about such complex reactions, for this "systems" approach to biochemistry is in many ways an area that is just being defined within molecular biology itself.

To get an idea of biocomputing in action, we can consider one of its founding, proof-of-concept experiments as our starting point: Leonard Adleman's 1994 experiment involving the use of DNA to solve a standard directed Hamiltonian path problem. In doing this, we will pay particular attention to the techniques and technologies employed in the articulation of this DNA computer. This attention to detail will be a way for us to highlight the implicit claims that are materialized in biocomputing practices. From there we can then go on to consider some of the more theoretical implications of biocomputing, especially in relation to the issue of "computability" and the notion of distributed networks.

Biological Solutions, Computational Problems

Leonard Adleman's 1994 paper "Molecular Computation of Solutions to Combinatorial Problems" is widely regarded as a seminal proof-of-concept paper on the possibility of computing with biomolecules. Although speculation as to the computational aspects of DNA and other biological components was not uncommon prior to Adleman's experiment, it was this paper that translated or "ported" the informatic models of the biomolecular body into computational terms.[17] Adleman's DNA computer is far from a fanciful mathematical exercise; its concerns are simultaneously biological, computational, and mathematical. Indeed, its relevance, from a philosophical perspective, is that it renders computational concerns inseparable from biological concerns.

In his paper, Adleman shows how a particularly difficult type of mathematical problem can be solved using DNA and the standard tools of molecular biology. Consider a transportation problem such as the "traveling-salesman problem." You are a salesman, and must pass through seven cities on your itinerary. You do not want to waste money on plane tickets going back and forth to cities you have already visited, so you want to find the most efficient means of hitting each of the seven cities. In addition, you are beginning at the first city (O_1 or O_{start}), and must end up at the seventh city (O_7 or O_{end}). Your problem is thus: what is the most direct, efficient path that begins at the first city, ends at the seventh city, and passes through each city only once? (Figure 11).

Such problems are often found in the field of mathematics known as "graph theory." Graph theory, as a branch of geometry, is primarily concerned with the quantitative properties of networks. A "graph" (G) is therefore a set of "nodes" or "vertices" (V), some of which may be connected by "links" or "edges" (E). Nodes can be people, cities, or viruses, and edges can be social interaction, transportation, or infection. A given set of nodes can have a radically different structure (or network topology), depending on how the edges are configured, just as the linking capability of edges is directly related to the configuration of nodes.[18]

The traveling-salesman problem is easily solvable on the average personal computer, but with one important requirement: that the input data is very small. A network such as the one described here, with a mere seven cities, is even solvable by visual inspection and pencil and paper. However, as the input data size increases, the possible solutions to the problem also increase. If we take a simplified version of three cities (nodes), there are six possible routes from any given node: $3 \times 2 \times 1 = 6$ (or, in mathematical terms, 3!). What about ten cities? The mere addition of seven cities to the network would seem to matter little. But the possible number of routes increases

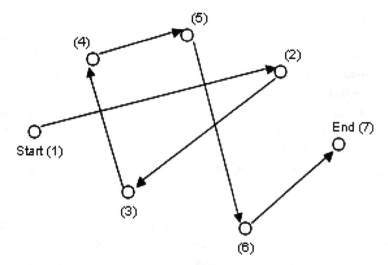

Figure 11. Diagram of solution to directed Hamiltonian path problem. Nodes represent cities, lines represent paths (adapted from Adleman).

to 3,628,800 (or 10!). The math is the same, but, as the input data size increases linearly, the search field for a solution increases exponentially.

This class of problems, known as "NP-complete," is characterized by its exponentially large search field. This means that, in some cases, there is no way to know if a solution to a problem exists, other than simply trying every possible solution. Note that the issue with such NP-complete network optimization problems is not finding a solution, it is merely deciding whether or not a solution exists at all. The key threshold in NP-complete problems, such as the traveling-salesman problem, is therefore in the efficiency of a given algorithm for finding a solution. In computational terms, this means considering whether a computer can find a solution in a reasonable amount of time. As the search field increases exponentially, so does the time required to find a solution.

The problem, though it deals with transportation (cities connected by highways or flight paths), could just as easily be considered in terms of communications (data routed through computers) or even epidemiology (infection and spread); that is, the traveling-salesman problem is an example of a network optimization problem, in which, for a given network, a certain value must be minimized. The network exists less as a spatial configuration, and more as a time-based actualization of a path through the network. In graph theory terms, the traveling-salesman problem is a version of a "directed Hamiltonian path" problem. It is "directed" because the edges of the graph are unidirectional (no backtracking; having one-way streets). It is "Hamiltonian" because it is named after the Irish mathematician William Hamilton, who devised a mathematical game involving tracing single paths along complex geometrical shapes. And it is a "path" because it does not repeat any node (in contrast to a "walk," which may repeat nodes, and a "trail," which does not repeat any edge).[19] The directed Hamiltonian path specifies a very particular way in which a network is realized, by actualizing some edges/links, while leaving others in an unactualized state.

The important thing to relate concerning graphs/networks and graph theory is that the node/edge distinction often follows a space/time distinction; that is, nodes are often taken as being relatively static (e.g., cities on a transportation route), whereas edges are often taken as being dynamic (e.g., moving from one city to another via plane). We will return to this observation later, but for the time being it serves to denote a particular type of ontology concerning physical systems (be they computer networks or biological networks).

The problem Adleman sets out to solve using DNA is thus a directed Hamiltonian path, or a specified version of the traveling-salesman problem. Because Adleman chose a small network (seven nodes), the solution is easily verifiable. However, Adleman's experiment is not intended to be mathematical research, but rather to demonstrate the feasibility of computation at the biomolecular level. Adleman's basic algorithm for solving the traveling-salesman problem follows four basic steps: (1) randomly generate a series of paths through the network; (2) keep only those paths that have the correct starting and ending points (V_{in} and V_{out}); (3) keep only those paths with the correct number of nodes (seven nodes or cities); (4) keep only those paths that hit each node once. As can be seen, the algorithm begins by generating a large group of possible solutions, then proceeds by a series of filtering processes, by first identifying starting/ending points, then the number of nodes in the path, and finally the identification of nodes in the solution paths.

The key biomolecular process that enables Adleman's experiment is the basic binding characteristics of DNA. As is well known, DNA, or deoxyribonucleic acid, is a double-helical structure, whose strands are composed of a sugar, a phosphate, and one of four nitrogenous bases (adenine, cytosine, thymine, guanine). Although DNA is as structural as a molecule as it is informatic, it has long been a tradition to "read" DNA as a linear sequence, thus making it amenable to treatment as a data string (where the beads composing the string are the variations of the bases). The way in which these variations of molecules bind to each other is widely acknowledged to follow regular rules. Known as "Watson–Crick base pair complementarity," it simply states that the binding affinities of DNA follow a regular pattern, in which adenine (A) always binds to thymine (T), and cytosine (C) always binds to guanine (G).

Under certain conditions of heating and cooling, the double helix of DNA can be broken apart ("denaturing"), synthesized ("replication"), and glued together ("annealing"), a process that has in fact been automated in the technology of PCR (polymerase chain reaction). From this basic principle of base pair complementarity, DNA contains two elements crucial to any computer: a processing unit (the enzymes that denature, replicate, and anneal DNA), and a storage unit (the regulatory "instructions" encoded in DNA strings). Not only does DNA form a highly efficient storage system (an estimated one bit per cubic nanometer), but, in the living cell, instructions are carried out in a massively parallel fashion (in contrast to the sequential processing of instructions in many computers). This combination of massive parallelism and storage capacity makes DNA an ideal "computer" for instructions that require simple calculations on a massive scale—problems such as the traveling-salesman or directed Hamiltonian path problem.

To illustrate the way in which Adleman's experiment layers computational and biological concerns, it will be helpful to describe the way in which the individual steps of the algorithm were carried out.

The first step was to generate random paths through the network. Because the Adleman experiment works with DNA as its "hardware," the first step involved encoding each of the seven nodes (cities) into DNA. Adleman chose to encode each node as a short segment of DNA (20 base pairs, or 20bp). Each node can therefore be represented by O_n, where n is between one and seven. Then each edge or link was encoded into DNA as well. However, because the "edge" DNA would connect two "node" DNA, the edge DNA would therefore be half of each node DNA. For a node O_3 and a node O_4, the edge connecting them would be half of each, or O_{3-4} (Figure 12).

Taking advantage of DNA's base pair complementarity, Adleman could then produce equal parts of single-stranded nodes (actually, the complementary strand of the nodes) and single-stranded edges. These "sticky" single strands of DNA could then be mixed together and allowed to bind accordingly. This mixture produces, through the combinations of possible binding between DNA strands, a large pool of DNA strands that represent possible solutions to the traveling-salesman problem (Figure 13).

The second step is to begin filtering this data set by keeping only those paths that begin and end with O_1 and O_7, respectively. This can be achieved using PCR, a kind of DNA Xerox machine, which repeatedly heats and cools DNA while adding extra DNA fragments, or "primers," to synthesize copies of DNA in the process.[20] In this case, the selection of primers is key. Primers that contain O_1 and O_7 will therefore replicate only those paths that either begin or end with O_1 or O_7 (note: this does not yet concern the length of the paths, or their order, only their beginning and ending sequences).

The third step in the algorithm is to filter again the paths retained from step 2. In this filtering step, only those paths that contain the exact number of nodes in the path are kept. Because each node was encoded with a DNA strand of 20bp (twenty base pairs long), any possible solution to the traveling-salesman problem will be exactly 140bp (20bp × 7 cities). Such a filtering process can be carried out using a technique known as gel electrophoresis, commonly used to sequence DNA.[21] In short, DNA strands to be analyzed are passed through an agarose gel using electric polarization (because DNA has a negative charge, it will be pulled by the application of a positive charge). The gel forms a thick mesh through which short strands of DNA will pass faster than longer strands. Using this technique, researchers can measure the length of individual DNA strands, using DNA of known lengths as "controls."

Node	Code
O_2	TATCGGATCG GTATATCCGA
O_3	GCTATTCGAG CTTAAAGCTA
O_4	GGCTAGGTAC CAGCATGCTT
$\hat{O}_3$	CGATAAGCTC GAATTTCGAT

Edge	Code
$O_{2\text{--}3}$	GTATATCCGA GCTATTCGAG
$O_{3\text{--}4}$	CTTAAAGCTA GGCTAGGTAC

Figure 12. Each node (city) and edge (path between cities) encoded as a DNA strand (adapted from Adleman).

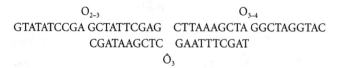

Figure 13. Two "links" from a single node, showing overlapping binding by base-pair complementarity (adapted from Adleman).

The results from the gel electrophoresis will therefore keep those paths of exactly 140bp (seven nodes). From this, the fourth step provides the last filter, that is to keep those paths that hit each node only once. This is where the directed Hamiltonian path qualification comes in, requiring that the path is not only directed, but that it does not repeat itself. A version of PCR is here used ("graduated PCR") in which selected DNA fragments are attached to a magnetic bead, which "tags" the selected DNA fragment (e.g., the DNA strand for node O_4). Using this technique, individual nodes along the path can be identified (e.g., with graduated PCR with O_4 as the primer, one would look for results that are 4×20bp or 80bp long). Repeating this process for each node will eventually filter out any paths that visit any single node more than once. If there are paths retained after this, they will constitute a solution to the traveling-salesman problem, because they will be directed (starting at O_1 and ending at O_7) and will visit each node only once (will constitute a "path").

Working backwards, we can see how each step of the Adleman experiment combines computational and biological elements in a common way. At each step, a common technique in molecular lab biology is used to carry out a "calculation": DNA synthesis performs encoding procedures for the input data set, gel electrophoresis performs data analysis based on the physical and electrical properties of DNA, and PCR performs data analysis based on identifying nodes in a combinatorial fashion. Again, the key element that enables calculation to occur at all is the principle of base pair complementarity in DNA.

We should note the obvious: though Adleman's primary concern is computational, his materials are almost totally biological. One would at first be tempted to state that this is a paradoxical notion of computer technology, because there is no technology present, only the biology of DNA and enzymatic reactions. However, it is just this initial presupposition that makes biocomputing worth noting as an instance of "biomedia." Certainly, there is no "technology" in this computer, if by this we mean electronic digital computers based on integrated-circuit technology. But in every other sense, Adleman's biocomputer is thoroughly technological, its "natural" hardware notwithstanding. It is a kind of technologization of biology, in which biology is technically recontextualized in specific nonbiological ways.

Thus, Adleman's DNA computer makes computing and biology inseparable, but in a certain way. It integrates the logic of the modern computer (input/output, memory, processor, logic blocks) with the structural properties of biological components and processes. This is more than a simple grafting of Boolean operators onto the physical medium of DNA, for, as Adleman and other biocomputing researchers note, DNA is a unique "computer" in its own right. Its parallel processing, dual binary logic, and immense storage capacity make for an entirely novel computing system.

"So please just send in the machines"

As Adleman's experiment makes clear, the biocomputer is not simply a microelectronic black box; rather, the "object" of the biocomputer is a clustering of techniques from genetic engineering and molecular biology, discrete mathematics (particularly graph theory and combinatorics), theoretical computer science (particularly concerning the question of "computability"), and the still-vague area of molecular or nanocomputing. Biocomputing, in the Adleman model, is therefore as much biological as it is computational.

In this sense, several historical marking points are worth pointing out in relation to biocomputing. Using Adleman's experiment as our paradigm, we can outline a series of conditions that conceptually enable the field of biocomputing.

There is, first, a series of statements that are taken either as assumptions or as statements actually implicitly made by biocomputing research. As a particular intersection between molecular biology and computer science, biocomputing begins from an assumption concerning the equivalency between genetic and computer "codes." This relationship itself emerges from the historical backdrop of postwar developments in both cybernetics/information theory and molecular biology. The discursive transactions between these fields in part informs Francis Crick's concern throughout the 1950s and 1960s with "the coding problem" in DNA.[22] The research of Crick and others into "cracking the genetic code" therefore supplies an interdisciplinary backdrop against which the questions of biocomputing are posed.

Yet, despite the basic tenet of this backdrop—that the DNA molecule in the living organism is organized and even functions like a "code"—biocomputing still needs another statement to make available the question of computation with respect to biological systems. For molecular biology, the trope of the genetic "code" is in itself enough to generate research into later efforts in genome mapping and genetic engineering. For biocomputing, by contrast, a further statement is required, and that is supplied by the studies in molecular biology on genetic regulation. François Jacob and Jacques Monod's well-known work on "genetic regulatory mechanisms" in the cell—resulting in the identification of "repressor" and "promoter" proteins and genes that turn other genes off or on—establishes a view of the cell that in many ways moves beyond the trope of the genetic "code."[23] Jacob and Monod's work suggests, using their own words, that the genetic material functions as a kind of "genetic program," regulating and modulating the states of other genes, which in turn dictates which proteins are produced and which are not. The trope of the genetic program obviously suggests that the extension of concepts from cybernetics into molecular biology will reconfigure the living cell as a cybernetics system, complete with feedback loops, regulatory mechanisms, and even subroutines. The reason Jacob and Monod's work is relevant to biocomputing is that it provides an early instance in which the living, functioning cell is possibly regarded as more than just a biological system; it is also a program, a computer in its own right, with its own type of formal logic (which remained elusive to Jacob and Monod, despite their use of terms from computer science and cybernetics). The work of Jacob and Monod on genetic regulation is an important statement concerning the nontropic quality of the notion that DNA is a computer. Although this is not explicitly stated by Jacob or Monod, the issue of the computability of DNA is raised here in an early form.

These two statements—that DNA is information, a "code," and that DNA is a computer, a "program"—can be seen to have created a fork in the road for future developments in biotechnology generally. One path is a familiar one, grounded in molecular biology, medical biotechnology, and genetic engineering. It is the path that extends the tropic quality of molecular biology, and from that begins to inquire into the biological applications of computing, leading to contemporary fields such as bioinformatics and genomics. It broadly goes under the rubric "computational biology". The other path is less familiar, and more recent. Instead of extending the tropic qualities of molecular biology, it takes molecular biology quite literally, and approaches biomolecules, cells, and interactions as being computers in their own right, related to but different from silicon-based, microelectronic computers. This "biological computing" (or biocomputing for short) is less concerned with the biological, biomedical implications of the informatic tropes in molecular biology, and is much more concerned with the computational, mathematical implications in biological components and processes. It is important to note that in neither case is informatics simply taken as a "metaphor" for something else that is material. Rather, both computational biology (the biological strand) and biological computing (the computer strand) ceaselessly materialize the relationship between biology and technology, genetics and informatics. However, they do so in different ways, and that difference is dictated in part by their approach to the organism. From the perspective of computational biology (i.e., bioinformatics), the organism is an information-processing system that constantly correlates sequence and structure (DNA and protein) in the ongoing functioning of the living cell. From the perspective of biological computing (i.e., Adleman's DNA computer), the organism is not the question; the question is to what degree isolated biological components and processes may form highly specialized, "wet" computers. The question that biocomputing puts to the organism is, "Does the question of 'computability' apply to the organism?"

One of the primary propositions that biocomputing makes is that biological components and processes contain within them a set of characteristics that lends them to problems of "computability." In short, the basic proposition of experiments such as Adleman's is that DNA is a computer. Again, it is worth restating the difference between this statement and the statements made by molecular biology (including genetics and biotech generally), that DNA functions in an informatic manner. In the latter statement (DNA operates informatically), the reference point is the notion of "information" derived from cybernetics and information theory. DNA is seen to operate biologically, but, in doing so, it is also seen to fulfill the requirements of technical systems (i.e., communications and control systems in engineering). By contrast, the former statement (DNA is a computer), made by biocomputing, takes as its reference point the notion of a computer derived from the theoretical computer science work of John von Neumann and Alan Turing. It does not take cybernetics/information theory as its standard, against which to analyze molecular biological interactions; rather, it adopts the theoretical approach of computer science, asking not whether DNA fulfills the definitions of information, but whether DNA might be a specific, unique type of computer in its own right. Certainly, there is a technical notion of information involved in this inquiry, but rather than grafting a series of technical requirements upon DNA, the biocomputing approach—arguably the approach of theoretical computer science—poses a set of challenges to what may possibly be a feasible, totally biological computer system.

Turing's work in theoretical computer science is especially interesting in this context, for his basic concerns lie at the heart of the contemporary inquiries into biocomputing. Turing's early work in the mathematical problems of computability begins from an interest in the paradoxes inherent in formal logic (some might say the paradoxes produced by formal logic). Spurred on by the work of Kurt Gödel (and later, David Hilbert) into the issue of mathematical validity, Turing's paper "On Computable Numbers," published in 1936, set out to articulate the mathematical limitations or domains of activity for calculation.[24] The challenges put forth by mathematicians such as Gödel were based on the internal contradictions that formal logic sometimes generated. For example, take the statement "This statement is false." If it is proved to be true, then the statement is no longer false (it leads to contradiction). If, on the other hand, it is proved to be false, then the statement is no longer true (in that it denies its own validity). Gödel suggested that such logical conundrums posed a challenge to mathematics, for they showed that such inconsistent problems were in some way representative of mathematics itself. Even taken modestly, the implication of Gödel's work was that an axiomatic system such as mathematics in some senses cannot help but to make statements that are arbitrary (that cannot be proved or disproved). Although Gödel's work generated much controversy within the world of mathematics, critics often pointed out that, despite such inconsistencies, buildings and bridges still stood, implying that Gödel's work was of only theoretical interest, and that such inconsistencies were anomalous by-products of a self-consistent essence of mathematics.[25]

For many, it was Turing who showed that the work of Gödel was of both theoretical and practical interest. Turing posed the question not of mathematical validity or certitude, but rather of the possibility of identification (or nonidentification) of such arbitrary statements. Rather than pore over the aporias of mathematical certitude, Turing focused on what is often called "computability": how can the difference between mathematically certain and mathematically arbitrary statements be identified? Turing also framed his question in terms of reflexivity: could arbitrary statements be identified from within the system itself? If a set of basic rules for differentiating between mathematically certain and mathematically arbitrary statements could be identified, then, in theory, this algorithm could be carried out by a machine, so as to automate the filtering process for practical applications in engineering.

Turing devised a scenario involving a "universal Turing machine," a machine capable of functioning like any other machine, once the instructions for that machine are fed into it. This universal computer could therefore simulate any other computer, whether its task was to calculate tables, send data, or play chess. The challenge for this machine was now to ask what would happen when it was fed its own instructions—that is, when it was asked to simulate itself. Such a problem would not be computable for Turing, for the computer would end up in a kind of digital psychosis, endlessly turning in on itself with no verifiable output. In other words, the Turing computer would be performing the kind of logical paradox articulated by Gödel earlier. From Turing's thought experiment, "computability" came to be defined in relatively precise terms: the situation in which it was possible to formulate a set of rules to decide whether or not a statement was susceptible to proof using only the rules of that system itself.

What does Turing's work on computability have to contribute to biocomputing? For one, it raises the question Turing and others posed at the early stages of electronic digital computing: how can a physical system be designed that would fulfill the mathematical requirements for computability? That is, how can you build a computer machine that would consistently function in a mathematically valid manner? The experiments of biocomputing researchers such as Adleman have suggested that biological computer systems can only be verified by the kinds of proof-of-concept experiments that rely on a presumed solvability of a given problem (i.e., the limited, easily solvable scope of the seven-node Hamiltonian path problem).

If we shift our perspective for a moment from computer science to molecular biology, we can put Turing's question another way: if Turing's computability thesis does establish a foundation for the level of validity in a calculating or computing system, and, if such systems can be constructed from biological as well as electrical components, then it follows that biological systems—as computing systems—inherently contain a set of logical inconsistencies that may be termed "noncomputable." In other words, if we accept the feasibility of biological computers, then we are also impelled to ask whether biological systems display some level of paradox, arbitrariness, or "unsolvability." This is tantamount to asking, in molecular biology, whether biological systems can, in certain cases, continue to function in an anomalous manner. Note that this is not asking whether or not a mechanical computer or biological system can malfunction or "break down" (systems crash or death). Rather, it is asking whether computers or biological systems can take on anomalous modes of functioning that, in their own right, are perfectly consistent, but that, from the vantage point of the computer or biological system, are identifiable as anomalous.[26]

DNA Dualisms

While the very idea of DNA computers raises the questions Turing posed concerning "computability," they also lead to a set of more intricate questions concerning the status of DNA computers as living organisms. The question that many designs in biocomputing elicit is this: if a computer can be constructed from biological components and processes, to what degree is this computer "living"?

To address this question, we can briefly consider the correlative within computer science: the notion of "intelligent machines" and its related discourses (AI [artificial intelligence] and branches of philosophy of mind).[27] As a way of comparing organisms and machines, as well as humans and computers, the discourse of intelligent machines attempts to discover the limits of computers, as well as the illusions of anthropocentrism, or the intractable uniqueness of the human being-as-organism. Although this tendency to compare the organism and machine is not new, it takes a specific form in the development of the modern digital computer during World War II and the Cold War era. In particular, the work in theoretical computer science of Alan Turing and John von Neumann provides us with two paradigmatic ways in which organisms and machines have been related in terms of the computer.

The first approach is illustrated by the theoretical and mathematical work of von Neumann, elaborated in a series of lectures given at Yale University in the mid-1950s. However, as an adviser to many of the military-sponsored mainframe computing projects (ENIAC and EDVAC), von Neumann's interest in the relationships between humans and computers dates back to the mid-1940s and the development of the "stored-program computer."[28] Von Neumann was an adviser to the building of the ENIAC, perhaps the most well known of the room-sized mainframe computers. In his collaboration with ENIAC researchers, von Neumann helped to develop the notion of a computer that had inputs, outputs, a processor, and a central memory unit. This logical structure, or "von Neumann architecture," remains, with several modifications, in the computers of today, and the very language of digital computers—processor, memory, and bits—derives in large part from von Neumann's design. Rather than separating program and data, the computer's memory could treat both as data, sidestepping the laborious process of plugging and unplugging wires for each procedure.

The von Neumann architecture contributes two important elements to our consideration of the relation between humans and computers. The first is the design itself. In brief, the von Neumann architecture is made of five components, each with a particular function in the whole: input and output units, a control unit, an arithmetic unit, and a memory. The program instructions and data on which the program operates are stored in the memory, while the control unit interprets the instructions, and the arithmetic unit completes the actual calculations. The inputs allow program instructions and program data to be fed into the memory, while the outputs display the results of the control and arithmetic units (one important change in current digital computers is that the control and arithmetic units are often considered together as the processor). The second contribution relevant to human–computer discourse is von Neumann's suggestion that binary rather than decimal encoding schemes could be used to represent data. Whereas the convention at the time was to use ten "flip-flops" of a switch to represent a single digit, the use of binary encoding schemes would enable three digits to be represented, owing to the combinatory possibilities (ten flip-flops of the switch, each flip encoding a different combination of values).

Taken together, these two innovations in modern computer design not only have had a technical impact on future computer science thinking, but they have also, in part, set the terms in which a comparison between humans and computers could take place. The articulation of "computability" into distinct sectors of the computer has an effect on how we think about the "processing" of data into knowledge, and in this sense the von Neumann architecture is perhaps more revealing as a mirror of human cognition than of the operation of a computer. This segmenting of the computer is therefore not unrelated to a segmenting of the processes of cognition, a comparison made more explicit in the field of cybernetics. This, combined with the implementation of binary coding and the use of "bits," makes for a view of the computer not only as segmented, but as combinatoric. The use of binary coding schemes allows for a greater flexibility in the storage capacity of the computer, while also allowing the encoding of all data—including programs themselves—into the computer's memory. In a sense, then, the first aspect of the von Neumann architecture (the five components) serves a segmentary, canalizing function, a kind of bureaucratic division of labor among interrelated component parts (first input into memory, then instructions to the control unit, etc.). If this is the case, the second aspect of the von Neumann architecture (binary coding) works against the first, allowing for a diversification of how much and what kinds of data can be stored (data in will be transformed into data out). While the first aspect "stratifies" or segments computer function, the second aspect "smooths" out the plane of what can be encoded (including the very description of the computer itself, which leads to Turing's notion of a "universal machine").

In addition, it is noteworthy that, owing largely to von Neumann's interest in neuroscience, terms such as *memory* replaced traditional terms such as *storage*. Von Neumann's lectures on the computer and the brain explicitly explore the technical significations of this use of biological language to name computational processes. This interest was specifically a technical, functional interest. Rarely does von Neumann speculate on the capacity for computers to obtain intelligence, sentience, or anything related to "mind." Rather, his interest lies in the low-level processes of how data is input, interpreted, operated upon (calculated), and output toward certain action. Von Neumann systematically provides technical descriptions of modern digital computers alongside findings in neuroscience and studies of the brain. His language pushes the possible resonance between computer and brain through the use of physiological terms:

> The organization of large digital machines are [sic] more complex [than analog computers]. They are made up of "active" organs and of organs serving "memory" functions—I will include among the latter the "input" and "output" organs.[29]

In this analysis, von Neumann notes that while modern digital computers utilize the segmentation and binary coding of the von Neumann architecture, the brain contains a combination of both analog and digital components. Interestingly enough, von Neumann locates the primary example of this in the activity of neurons, as well as in the genetic basis of nerve cells. Neurons contracting a muscle, or genes directing the synthesis of proteins, are examples of hybrid analog-digital systems in the body: "processes which go through the nervous system may . . . change their character from digital to analog, and back to digital, etc., repeatedly."[30] While the firing of a neuron or expression of a gene may be viewed as digital, their outputs—muscle contraction or protein folding—may be understood as analog.

However, in spite of these differences, von Neumann's low-level, materialist approach tends more toward the correspondences between the computer and the brain. In a series of comparisons that were to prove influential for a certain branch of AI, von Neumann essentially views the brain—and neuroscience—through the lens of modern digital computers:

> systems of nerve cells, which stimulate each other in various possible cyclical ways, also constitute memories . . . In our computing machine technology such memories are in frequent and significant use . . . In vacuum-tube machines the "flip-flops," i.e. pairs of vacuum tubes that are mutually gating and controlling each other, represent this type. Transistor technology, as well as practically every other form of high-speed electronic technology, permit and indeed call for the use of flip-flop like subassemblies, and these can be used as memory elements in the same way.[31]

In other words, it is highly relevant that von Neumann's lectures were titled "the computer and the brain" and not "the computer and the mind." Despite this, we can detect in von Neumann's analysis an ambient interest in mind as well as brain. The use of terms borrowed from biology, as well as the direct comparison of computer and brain (through the lens of computers), indicates that such a comparison is not without its at least implicit interest in the kind of "mind" that would arise from the brain—whether this brain is a neurobiological one, a vacuum-tube one, or a transistor one. Certainly, terms such as *memory* become quite contentious here, and von Neumann's analysis can be seen to consistently border the line between an emphasis on lower-level brain functions and higher-level manifestations of mind. The questions implicit in von Neumann's analysis are twofold: First, on what level can the computer and the brain be seen as functionally analogous, and how might such a correspondence affect the design of technical systems? Second, if brain is not unrelated to mind, what "higher-level" manifestations might arise from the lower-level functions of computers?

To unpack the implications of these questions, we can turn to our second example of thinking concerning the relation between humans and computers: the work of Alan Turing and the famous "Turing test." Between his work for the British government's National Physics Laboratory in the late 1940s, and his lead on the MADAM computer project at Manchester University in the 1950s, Turing's interest in the theory of computation turned increasingly to questions of cognition and communication.[32] Turing's famous 1950 paper "Computing Machinery and Intelligence" set the terms for a still-ongoing debate about the feasibility of intelligent machines, as well as articulating the theoretical questions for the field of artificial intelligence.[33] One of Turing's insights was that the question of free will and determination in computers is, in a sense, a moot question. What counts is not whether or not a computer really is intelligent (and whether or not we can deduce valid criteria for assessing this), but whether or not a computer behaves as if it were intelligent. The implication here is that our own criteria for intelligence in humans (and other species, for that matter) is largely dependent on the ways in which we act and interact with others, and less dependent on any universal set of measurable criteria.[34] It is this precondition of interaction that gives the impression that the computer-as-player is somehow "intelligent" or "alive."

To demonstrate this further, Turing hypothesizes an experiment in which the "intelligence" of a computer can be assessed, but assessed in a way that would bypass the unending philosophical arguments of free will, sentience, and consciousness. This experiment—the "Turing test"—takes as its starting point communication between individuals separated physically from each other in separate rooms or by dividing walls.[35] Suppose that person A in one room communicates through a keyboard and terminal with person B in another room. Their communication is mediated by the monitors (that is, language as a cultural practice is the mediator). Depending on the conversation, person A may make assumptions about person B, about gender, ethnicity, age, personality type, and so on.[36] Now, Turing proposes, suppose we replace person B with a computer, programmed such that it is able to answer typed questions and, based on those questions, return responses and pose its own questions. As Turing states, "may not machines carry out something which ought to be described as thinking but which is very different from what a man does?"[37] Turing's point here is that if person A is in any way convinced or in any way assumes that there is a human being on the other end, and not a computer, then the computer will have effectively demonstrated a certain level of "intelligence." Why? Because—and this is Turing's most controversial point—this is how human beings assess "intelligence" in each other.

In the context of biocomputing, the Turing test is noteworthy, not because it presages anything in biocomputing, but precisely because it rules out the possibility of biocomputers. Note that Turing's question is not "is biology computational?" but rather "is intelligence computational?"[38] In other words, it might be more accurate to say that Turing relates organisms and machines in the specific terms of humans and computers. This is done on the level of a certain type of cognitive performance; humans and computers are compared via the notion of "mind." By contrast, von Neumann's interest in the functional correspondences between the computer and the brain expresses a different sort of question, which also forecloses the possibility of biocomputers. If Turing's question concerns "intelligence," von Neumann's question is, "Is memory-based cognition in the brain a stored-program computer?"

To simplify greatly, we can suggest that while Turing emphasizes intelligence through the performance of the Turing test, von Neumann emphasizes memory as data processing through the architecture of the stored-program computer. However, these terms need to be understood in very particular ways. For Turing, "intelligence" is not an a priori quality exclusive to human beings, but is rooted in communication, performance, and practical assessment of behavior. Likewise, for von Neumann, "memory" is not so much a mysterious set of impressions specific to human beings or other organisms, but, when seen functionally as a pattern arising from switches (flip-flops), it can be considered a mechanical property of biological or computational systems. At no point in their respective texts do Turing or von Neumann question the existence of intelligence or memory in human beings as we commonly use the terms. Although Turing, not without some irony, asks "why shouldn't I be considered a computer?" it is important to note that both Turing and von Neumann begin by analyzing cognition in the organism through the lens of computer science—for both, the data processing involved in mind, brain, and computers proceeds through a series of discrete, finite-state machines. This is evident in the Turing test (series of questions and answers) as well as in the von Neumann architecture (correlation of control unit and memory unit, of arithmetic unit and control unit). In the process, the terms themselves (intelligence, memory) become transformed, not only quantitatively, but also qualitatively. We might say that for Turing, intelligence is a "state" of mind, whereas for von Neumann, the brain is just the "memory" of the computer.[39]

The field of biocomputing offers a corrective to the assumptions in the intelligent machine discourse, but it also raises its own set of problematics not explored by either Turing or von Neumann. If Turing and von Neumann see the human–computer relationship in predominantly cognitive terms, gauging "the human" in terms of higher-level processes such as learning, memory retrieval, or communication, biocomputing sees the human–computer relationship in predominantly biomolecular terms, displacing any interest in the human with an interest in biomolecular process. Biocomputing inverts the intelligent-machine discourse's interest in cognition, and places higher priority on the seemingly secondary, lower-level processes of the organism at the biomolecular level. Biocomputing keeps the relationship of organism and machine at the level of organism and machine, and resists the analogous comparison of human and computer that both Turing and von Neumann carry out. The key to this difference is that for Turing and von Neumann, the differentiation between organism and machine takes place at the level of human cognition: intelligence/learning and memory/data processing are the limits of what computers can do.

By contrast, biocomputing suggests that the difference between organisms and machines is not anything human, but rather a difference between living and nonliving systems: cell metabolism, gene regulation, and cell membrane signaling are the limits of what computers can do. Again, we can detect not only a biologism but an anthropocentrism in Turing and von Neumann, in the sense that the human is the standard against which computer performance is judged. Biocomputing does not necessarily assume this; it looks rather to the complex, "parallel" processes in the living cell as the threshold of computability. For Turing and von Neumann, what is at stake is essentially mind, with the human as its most sophisticated manifestation (one that is nevertheless amenable to computation). For biocomputing, what is at stake is "life," by this meaning the ability of biomolecular systems to carry out exceedingly complex calculations "naturally." In a strange way, neither Turing nor von Neumann is really interested in computation, but rather the computational explanation of human-centered attributes such as intelligence, learning, or memory access. Biocomputing researchers, in contrast, are centrally concerned with computation, with the understanding that computation in the 1990s comes to take on more than it had in the 1950s. For biocomputing, computation becomes, in part, synonymous with complexity and parallelism. In this context, "life" is both nonhuman and "intelligent."

We can therefore see the changes in the way that the computer is related to the human as a shift from an emphasis on "mind" (or cognition) to an emphasis on "life" (or complex networks). The key link between this emphasis on mind versus life is the changing artifact of the computer itself. From a historical perspective, it is obvious that the computer shifts from a room-sized, military-funded "electronic brain" to a microelectronic, industry-marketed "personal" computer. Although the computer as an artifact plays many roles and takes on many meanings, the point to be made here is that, from the perspective of computer science, the modern digital computer of Turing and von Neumann conceives of computation as a cognitive function, whereas in the PC era of biocomputing research, computation is seen as inherently nonconscious, distributed, and in parallel.[40]

When Computers Were Human (Again)

Before concluding with a consideration of the role of networks in biocomputing, it will be helpful to provide a summary of our concerns thus far. As we have seen, the burgeoning field of biocomputing raises a number of issues pertaining to our discussion of the relationships between biology and computers. Looking at Leonard Adleman's initial DNA computing experiments, we saw how the algorithm for DNA computing of a computationally complex problem (such as NP-complete graph problems) made use of techniques and tools from molecular biology. In general, we can say that Adleman's DNA computing experiment shows us three things about the concept of the biocomputer. First, it provides a possible set of criteria for assessing what may count as "computational" for biocomputer systems. Second, it demonstrates the technical feasibility of designing biological systems that function in "nonbiological" ways. And third, the DNA computer materializes a network, both computationally and mathematically, as well as biochemically.

Thus, biocomputing not only rearticulates "technology," but in its procedures it also brings a unique perspective to the question of "computation." As a number of biocomputing researchers have suggested, Adleman's DNA computer is important because it poses the question that Turing posed in relation to mainframe computing: what qualifies as an adequate (solvable) problem for a computing machine? On the one hand, the historical-biological understanding of biological components (the cell, the genome, DNA) as functioning in an informatic way has been instrumental in creating the conditions for conceiving of DNA as a computer in itself. On the other hand, the creation of the DNA computer has also run counter to the common habit of aligning computers with either brain (in the von Neumann architecture) or "mind" (in the Turing test). We have suggested that one result of this rethinking of computational paradigms is the shift from "intelligence" to "life," or, to put it another way, from the neuronal to the metabolic model of network behavior.

This shift is noteworthy precisely because the assessment of computability in biocomputing takes place in the context of problems that are inherently networking problems. The graph-theoretical problem of the traveling salesman (directed Hamiltonian path) not only provides a test problem for DNA computing, but it also instantiates a network layering of biological and informatic networks. Thus, these first two characteristics of biocomputers—nonbiological instrumentalization and the question of computability—come together in a third characteristic, which has to do with the network dynamics of biocomputers.

As Adleman's research suggests, the application of DNA computers to NP-complete problems is a form of network computing. Not only does the DNA computer "solve" the Hamiltonian path problem through the analysis of DNA sequences, but it does so through a biological process in which multiple DNA "nodes" become linked or not linked to each other via a process of base pair binding (or an "edge"). There are, therefore, two network layers that are formed in Adleman's biocomputing algorithm. The first network layer is a biological one. This is evident from the very concept of the DNA computer as proposed by Adleman: a basic property of DNA in the living organism is its specificity in the binding of its base pairs (A-T; C-G). This biological property forms the "processor" of the DNA computer, in that it is the component that actually does the work of computing. The precomputing procedures of encoding (sample preparation input), as well as the postcomputing procedures of decoding (DNA sequencing output), are both dependent on the central processing unit (CPU) of the biological property of Watson–Crick base pair complementarity.

The second network layer is a computational one, a network not present in the DNA itself, but selectively extracted from a technique of decoding the double-stranded DNA sequences. The difference between the computational and biological layers is clearer if we consider two contexts. For the molecular biologist, the sequence would be decoded into a linear string of As, Ts, Cs, and Gs, and then imported into a range of software tools and analyzed against biological databases for possible homologies or polymorphisms; that is, sequence is first and foremost what counts for the bioinformatician and biologist. By contrast, the research in biocomputing takes the DNA sequence not as a string of (biological) data, but as a linearized arrangement of a graph (or network) of distributed nodes and edges composed of DNA; that is, in contrast to the biologist, the computer scientist's goal is to extract a graph from a string, to redeploy the network from out of the sequence, to decode the distributed from the linear. The network of this computational layer is therefore derived from the biological functioning of the biocomputer, *though that biological functionality has no biological function.* It is worth pausing over this statement: the biocomputer displays biological functionality because it creates a context in which selected biological components and processes may occur. In the case of the Adleman experiment, these are DNA base pair binding properties (Watson–Crick complementarity). However, this functionality serves no biological function, directly or indirectly, and is therefore quite different from DNA in the living cell or in the molecular biology lab. The DNA computer does nothing to the DNA itself that might be seen as beneficial from a biological, organismic, or medical perspective. However, it still "works," and that is perhaps the central insight of biocomputing.

This division between biological and computational layers is, of course, not a decisive one; it is understood from the beginning that the use of DNA's properties are being recontextualized radically for, in this case, computer science (and not molecular biotechnology). Likewise, the networks that emerge from such computations are in no way inherent in the DNA computer itself; they are a result of the recontextualization that occurs in the biological layer of the network, as well as a result of the decoding procedures specific to the particular mathematical problem being computed.

That being said, it is worth pointing out that, from the vantage point of biocomputing, there is an isomorphism between the two layers. Between the multiple DNA fragments (or nodes) placed in a PCR thermal cycler to facilitate their binding (or edges) and the resultant directed Hamiltonian graph there is a correspondence based on the specific relationships between discrete nodes. On the biological layer of the network (DNA binding via PCR), we have a number of DNA fragments, some which code for nodes (i.e., "cities" on the Hamiltonian path) and some of which code for edges (i.e., "roads" between cities that make up the directed path). If we view these DNA fragments (nodes and edges) spatially, what we are presented with is a tangled mass of points and lines, some of which will connect, and some of which will not. The task of the DNA computer is to perform the "computation," which in this case is the routine binding of base pairs. The task after this, however, is to extract from this mass the particular set of points and lines that represent the most efficient solution to the problem. This is done, as we have seen, by a particular means of decoding the sequence so that it produces a network (from string/sequence to graph/network). The resultant graph therefore bears some relationship, however direct or indirect, to the particular set of interrelated DNA fragments in the PCR cycler. In other words, the way in which the DNA computer's processor works is to move along a three-step process: from a distributed set of DNA fragments (nodes and edges), to a linear DNA sequence (node-edge-node), to a graph or pathway with nodes distributed in space (the directed Hamiltonian graph). In this three-step "processing" of data, we go from a graph, to a string, to a graph again (or from network to sequence to network). It is this first and third state that are isomorphic, and in this sense biocomputing can be seen as a means of backtracking to find out what has happened in the interactions of biological components and processes.

One way of understanding this difference between these two networks is to pay more attention to the role of time. Studies of computer processors are all concerned with time, especially in "clocking" a processor's ability to handle computationally intensive tasks (such as image processing). However, by time we mean something different than processor clock time, though not unrelated to the notion of time as a quantifiable unit. Our question is, in the potential relationships that exist in the DNA computer before processing occurs, what kind of a network exists? One way of approaching such a question is to pursue what we mean by "potential" relationships that constitute a network. Surely, from a more technical point of view, a network only exists by being materialized in some way; otherwise the network "in potential" is simply a totally connected network (because every node is potentially connected to every other node). From the vantage point of DNA computing, a particular type of network is being searched for—in Adleman's case, one that fulfills certain requirements of the Hamiltonian path problem (being directed or one-way, beginning and ending nodes, passing through each node only once).

But this computational, mathematical network that is searched for must somehow pass through a set of biological relationships, relationships that constitute the "processor" of the DNA computer. There are actually two types of processing, two types of temporality, in the DNA computer. Just as we have two types of network layers in the DNA computer, we also have two correlative time-based modes as part of the DNA computer processor. The first type of time is the "binding time" of the molecular interactions in the PCR cycler, a biological time configured by base pair complementarity (in the Adleman example). Actually, we need to be cautious in calling this "biological time," because the PCR machine does not aim to replicate or simulate in vivo conditions. Rather, the PCR machine isolates a set of particular biomolecular interactions (annealing and denaturing of DNA strands), and, through cycles of heating and cooling, intensifies and speeds up the biomolecular interactions that participate in the replication of DNA strands. (It is for this reason that PCR is a standard laboratory technique, for it enables the mass replication of any DNA sample for analysis and experiment.) However, we can provisionally refer to the DNA computer's processor as operating through biological time in the sense that the PCR cycler isolates and intensifies biomolecular interactions that regularly take place in the living cell.

The second type of time is the "polynomial time" of the particular computation to be solved, a computational time that is a set of limitations placed on the biological binding time of the DNA computer processor (the PCR cycler). In Adleman's experiment, this polynomial time is set by the class of mathematical problem to be solved, a problem class that is defined by its exponentially large search field. As a mathematical class, the NP-complete problems thus place certain sets of computational constraints on the way in which the binding time is configured. In one sense, this is obvious, in that the PCR cycler will be configured so that it optimally accommodates the kinds of results that the polynomial time requires (i.e., minimal cycling to enable a path sequence to anneal that matches the predicted outcome for the five "cities" or nodes of the problem). In this way, the polynomial time is not a separate process from the binding time of the PCR cycler. The processor of the DNA computer encompasses both, though in different ways. While the binding time dictates a set of possible biomolecular interactions that constitute the "engine" of the processor, the polynomial time nuances the binding time by setting a series of generalized result parameters (i.e., a path made of seven nodes, with a degree of two, with specific beginning and endpoint nodes). Just as the computer network layer is grafted onto the biological network layer, so the computational polynomial time is enfolded onto the biological binding time in the DNA computer processor. As time-based processes, one does not happen before or after the other; rather, their concurrence (of processor capacity and computability) defines an integrated temporality that is defined by the twofold constraint of biology and computability.

In a sense, all models for biocomputing—membrane, tiling, signaling—configure their processors according to this twofold, enfolded process of biological binding time and computational polynomial time. For Adleman's experiment—and arguably for biocomputing generally—the relation between the network preprocessing (binding time) and the network postprocessing (polynomial time) is that between the possible and real. A possible set of networks exist preprocessing, a possibility instantiated by the biological process of base pair binding. This possible network is therefore a network of combinatoric relationships. This network is negated in the postprocessing network in that, of those combinatorial possibilities, only one will be selected as the solution to the problem. The graph at the end of the experiment therefore simultaneously realizes a potential in the preprocessing network and negates that network as potential (because there can be only one).

Molecular Molecules?

If the biocomputer's processor functions via a binding time and a polynomial time, we can return to the question of network "layering" and ask: how are the computational and biological network "layers" coordinated in the DNA computer? That is, what might the isomorphisms between the biological and computational layers of the biocomputer tell us about the possible network dynamics common to both? This question pertains equally to philosophy and to computer science, for it brings together the questions concerning self-organization in biological systems with the questions concerning computability in biocomputers. What is needed here is a way of understanding the heterogeneity of networks in biological-computational systems such as biocomputers.

In a discussion of the common characteristics in packs of wolves, swarms of rats, transubstantiations in sorcery, the processes of memory, mineral "life," and epidemic contagion, Gilles Deleuze and Félix Guattari use the term *molecular* to describe dynamic changes in systems that exist above and below the level of the individual.[41] The concept of the molecular is defined not by its scale, but rather by its dimensions. The difference between a single wolf and a pack, between a single cell and a network of cells, is not simply number, but rather the modes of interaction that constitute a pack and an immune system as a network. The molecular is the phenomenon of intensity (aggregate dynamics) as opposed to extensity (movements of particles), qualitative transformations as opposed to quantitative analysis.

Central to the concept of the molecular are the concepts of becoming, difference, and multiplicity. As Deleuze and Guattari state, "the molecular is becoming," or the concept of the molecular implies dynamic transformation in organization. However, this notion of becoming is not be opposed to a notion of static "being." Rather than saying that a set of individual wolves becomes a pack, or a set of individual cells becomes an immunity network, we should say that there is becoming in each of the wolves, in each of the cells.[42]

Becoming is connected to two other concepts, difference and multiplicity. By difference, Deleuze and Guattari do not mean a negative notion of difference ("A is not B"), but rather a positive notion of difference that proceeds via a continuous internal restructuration in time ("A is not A"). This generative notion of difference creates novelty not by making anew, but by repeating itself, by repeating its internal differentiation ("I is another"). This principle of differentiation contributes to the third concept related to the molecular, that of multiplicity. As Deleuze and Guattari note, multiplicity can be both qualitative (a cluster, a bunch, a group, a pack, a swarm) and quantitative (clustering coefficients, power laws, graph topologies). Multiplicities are both singular ("a" pack, "a" swarm, "a" network) and plural (a pack of wolves, a swarm of insects, a network of cells).

Taken together, the concepts of becoming, difference, and multiplicity articulate the ways in which the molecular manifests itself in any network. The molecular can be described as a phenomenon that expresses the transformative dynamics of becoming, which is enabled by a principle of internal differentiation, which proliferates differences over time to create multiplicities.

In our consideration of biocomputing, we have seen several networks in action: the networks of biological components and processes in the living cell, the networks of interactions of those components in lab technologies (synthesizers, PCR, electrophoresis), the networks described by mathematical problems in graph theory (directed Hamiltonian path problems), and the abstract networks modeled by such mathematical problems (e.g., geographical networks of cities in the traveling-salesman problem, or information networks in routing Internet data). When we look at biocomputing as an instance of "the molecular," how does our understanding of these different networks change?

From a philosophical perspective, the preprocessor network of the DNA computer (the mass of DNA fragments in the PCR cycler) is an instance of a virtual network, in that the combinatoric characteristics of the network exist not to solve any one particular problem, but rather to continuously differentiate, and thereby establish connections between nodes. Furthermore, if we recall that the coding scheme of nodes and edges has been embedded into the DNA fragments (with no functional alteration to the DNA), then the type of network formed in the preprocessing state is one of "edges without nodes."

This network of edges-without-nodes is at first nonsensical, but also, in another way, rigorously empirical. How can a network exist without nodes? Don't the nodes precede the edges? Aren't they in fact a necessary condition for the action or movement of edges to exist? In any given network—the Internet, a social network, a metabolic network, an economic network—it seems to make sense that the discrete objects we call nodes (computers, people, cells, corporations) exist prior to the actions they effect (data transfer, enacting of dialogue, enzymatic reactions, fluctuations in value). If there is no subject, how can there be any action, any intentionality, any "directed" links between subjects?

On the other hand, it is equally clear that a static network is not a network at all. A group of computers can have cables attached between them, but if there is no activity on individual computers or between them, a network exists only hypothetically. Networks are materialized by their actions, by the edges that create proximities, alliances, and condensations between discrete nodes. In a sense, we can say that, although networks seem to depend on the preexistence of nodes, they also are constituted as networks by the edges between nodes. In the cell, a network without edges is death, just as a network with only edges is metastasis. Networks are always active and activated; they are always relations whose terms may very well change over time.

If this is the case, then it would make sense to reconsider biocomputing in light of a concept of edges-without-nodes. As already noted, biocomputing constructs two layers to its unique bioinformatic protocol: a computational layer, represented by the directed Hamiltonian path, and a biological layer, represented by Watson–Crick base pair complementarity. The molecular biology techniques of DNA synthesis, PCR, and gel electrophoresis provide the media through which one layer touches the other. What is the difference between these two layers in biocomputing? We can begin by suggesting that the computational layer, informed as it is by modern graph theory, is a network of discrete dynamics. This is illustrated in the graphical representation of the traveling-salesman problem (directed Hamiltonian path) (Figure 11), as it is illustrated in other graph-theoretical depictions that utilize a diagrammatic language of nodes and edges. In other words, the computational layer, working from a basis in graph theory, necessitates a topology of clearly demarcated nodes (static entities) and edges (effected actions). In this way, the computational layer is "computational" according to a digital, binary logic of Boolean operators (AND, OR, XOR).[43]

By contrast, the biological layer is "computational" in an analog manner, not unlike models of physical computers based on water, mercury, or mechanical parts. The various biological models for biocomputing—from DNA base pair binding to protein-protein structural specificity—are all based on processes that are continuous. Note that such processes can be interpreted as being discrete (e.g., genes being switched "on" or "off" in gene expression), but there is no separation between the medium and the message in the living cell. Indeed, this has been the primary technical challenge for both bioinformatics and biocomputing: to integrate the digital/computational and analog/biological layers of computability into a single whole.

If, from a philosophical standpoint, the main challenge for biocomputing is to functionally correlate its two protocological layers (the computational and biological layers), then we can suggest that Deleuze and Guattari's notion of "the molecular" can serve to prompt new modes of thinking in the understanding of how computers and biology mutually implicate and transform each other. The model of biological process in traditional molecular biology and biochemistry is one in which molecules are nodes, and the physical-chemical interactions between them constitute edges.[44] In other words, the view of biomolecular networks in biology and biochemistry is one based on the graph-theoretical model of discrete nodes that precede the effectuation of edges between them—the same digital model of computation seen in biocomputing's computational layer. But we also see that there are important differences between digital computers and living cells from a network perspective. Although one can be reduced to the other, it would also seem important to consider a notion of networks—in particular, biomolecular, biocomputational networks—that begins from the dynamic perspective of edges, rather than the static perspective of nodes. Experiments in biocomputing such as Adleman's show that computational problems can be solved, but that such novel, hybrid systems also generate networks at other levels as well. In the case of biocomputing, "molecules" are not necessarily "molecular."

CHAPTER FIVE

Nanomedicine

Molecules That Matter

Small Body Problems

To begin with, we can discuss the large effects of the very small. Linda Nagata's science-fiction novel *The Bohr Maker* envisions a near future in which the ability to control individual atoms has become a reality, a primary theme in the emerging scientific field of nanotechnology.[1] Although a number of science-fiction works take up this nanotechnological theme of control on the atomic scale, *The Bohr Maker* is notable because of the way in which it provides a specific meditation on how the human body may be transformed by nanotechnology.[2]

The Bohr Maker takes place in a world defined (and redefined) through nanotechnologies and socioeconomic divisions enhanced by the access to the design of those technologies. Although the future world of *The Bohr Maker* is divided into a first-world Commonwealth and a predominantly Southeast Asian third world, "citizenship" and even subjectivity itself are defined by nanotechnologies at the level of the everyday. In the novel, Phousita, a young Indian woman living in the slums of Sunda, comes into possession of a device called a "Maker." Unbeknownst to her, this device has been smuggled out of government research labs, where it has been kept under lock and key. The Maker is a nanotechnology device (a kind of nanotech PDA) invented by the scientist Leander Bohr. When injected into a person, it enables that person to modify certain biochemical features of the body, including genetically engineering one's own genome, enabling cellular and molecular regeneration (biological life extension), and enhancing neurological capacity (intelligence boosting). When Phousita first feels the effects of the Maker, she and her partner Arif attribute it to either a mysterious illness or even spirit possession. At this stage of the novel, Phousita's body is approached as an infected body.

However, Phousita eventually meets and falls in love with Sandor, who was involved in smuggling the Maker outside of the Commonwealth. Sandor and his brother Nikko have stolen the Maker in order to extend the preset life span of Nikko, a nano-biotechnologically engineered government "experiment." In the process, Phousita learns of the Maker's capacities, and her previously diseased/infected body becomes a nano-designed body, a body moving from the introduction of foreign material (infected, diseased, possessed) to a body that has incorporated new design affordances (internal biochemical modification).

Because Sandor and Nikko are on the run from the Commonwealth police, the group initiates a further bodily transformation in their flight from the police. They take refuge in cyberspace, by essentially replicating their biochemical and neurological data into cranial interfaces called "atriums." Sandor, Phousita's lover, must therefore "hide" in Phousita's atrium, not unlike a memory, a reactive memory. By replicating their nano-biotechnical makeup from the "real" world into the virtual world of information space, Sandor and Nikko use nanotechnology as a bridge between the atomic and the digital, between matter and information. This last transformation leads to the novel's conclusion, in which a special "biogenesis" function is enabled by Nikko and Sandor (in fleeing from the Commonwealth), a process of distributing their data that is simultaneously transcendence and death.

The Bohr Maker expresses several promises and also anxieties surrounding the ways in which our bodies may be transformed by fields such as nanotechnology. Implied from the novel's very beginning is an uneasy fusion between human and machine, flesh and data. Taking place in "the diamond age" of nanotechnology, we are never introduced to robots, cyborgs, or human–machine prosthetics. Instead, *The Bohr Maker* takes for granted a more nuanced relation between human and machine at the molecular-atomic level, in which the cell and the computer fold back onto each other, giving us quasi-organic devices such as the Maker.

In addition, *The Bohr Maker*, like other nanotechnology SF, gives another perspective on the "grey goo" problem that has been a topic of nanotechnology for some time. Put simply, the grey goo problem has to do with the restrained, uncontrolled replication of nanomachines, to the point where they reach epidemic proportions, possibly threatening ecosystems, populations, and human biology.[3] *The Bohr Maker* envisions the grey goo problem not as a horror story, but rather as a narrative of ambivalent emancipation, moving not in the direction of transcendence but in the direction of immanence.[4] Finally, *The Bohr Maker* raises the more immediate issue of the relationship of technological instrumentality to the human body, and indeed to matter itself. *The Bohr Maker* positions the Commonwealth as the site of instrumentalization, and the third world Phousita inhabits as the site of both underground individual self-fashioning and group emancipation.

However, nanotech SF such as *The Bohr Maker* leaves open the question of what it means to conceive of biology as design versus biology as engineering. The question of how nanotechnology may transform the very notions of instrumentality, technological determinism, and the relation between living and nonliving are all questions raised in nanotechnology research as well. Nanotechnology—the control of matter at the atomic level—is, like biotechnology, a research field that often promises a disease-free, biologically upgraded, posthuman future. But it is also a set of practices that may transform our notions of what it means to have a body, and to be a body. Nanotechnology works toward the general capability for molecular engineering to structure matter atom by atom, a "new industrial revolution," a view of technology that is highly specific and combinatoric, down to the atomic scale. But, in this process, nanotechnology also perturbs the seemingly self-evident boundary between the living and the nonliving (or between organic and nonorganic matter) through its radical reductionism. It is within this configuration of approaching matter as being "programmable" (or combinatorial) that we can begin to inquire into the ways in which the body as both living and nonliving matter is articulated within nanotechnology research.

Nanomedicine/Nanotechnology

Richard Feynman's famous 1959 paper "There's Plenty of Room at the Bottom" outlined a vision of future technologies that would be able to control and design a range of miniature devices at the molecular and atomic levels. Feynman speculated that, with the technological drive toward miniaturization, it was possible to build machines that would construct replicas of themselves at incrementally smaller scales, eventually reaching the level at which individual atoms could be controlled. (His examples included writing the *Encyclopedia Britannica* on the head of a pin, as well as tiny mechanical manipulators and computer storage devices.) As Feynman had pointed out:

> that enormous amounts of information can be carried in an exceedingly small space is, of course, well known to the biologists, and resolves the mystery which existed before we understood all this clearly, of how it could be that, in the tiniest cell, all of the information for the organization of a complex creature such as ourselves can be stored.[5]

However, the example from molecular biology provides us with models not only for information storage, but for acting on and through that information. Feynman continues, suggesting that, "biology is not simply writing information; it is doing something about it ... Many of the cells are very tiny, but they are very active." With a hopeful, even prophetic tone, he looks forward to a future where miniaturization makes possible what he calls "surgeons you can swallow."

This vision was a major influence in contemporary fields such as nanotechnology and molecular biotechnology. It also resonates with contemporary science fiction, which imagined the promises and anxieties of a near-future technology able to design and control matter—living and nonliving—at the molecular level.

However, it was not until the early 1980s that something resembling a defined discipline around this vision of engineering at the atomic level was formed. Eric Drexler's

Engines of Creation put forth an ambitious program, for what he variously called "molecular engineering" or nanotechnology. Put briefly, a nanometer is one billionth of a meter, or six carbon atoms wide. In contrast to what Drexler calls "bulk technology," or modes of technological production that handled matter en masse, a nanotechnology would work in a "bottom-up" fashion, focusing on the precision control of the individual atoms that compose matter.[6] Nanotechnology broadly aims to be able to control, engineer, and design matter at this level—the very building blocks of the material world, as it were. As Drexler points out, "molecules matter because matter is made of molecules, and everything from air to flesh to spacecraft is made of matter"[7] (Figure 14).

In *Engines of Creation,* Drexler outlines the types of nano-scale devices—or "nanomachines"—that would have to be constructed to realize this vision of molecular-atomic engineering. As we know, atoms interact with each other with varying forces and stabilities, composing larger aggregates, or molecules, which then function in diverse ways as organic or nonorganic matter. The technologies needed for design and engineering at this level would themselves operate at this level, somewhat like a tiny Tinkertoy construction. These include "assemblers" (nanomachines whose primary function is to assemble atoms into molecules), "replicators" (nanosystems whose goal is to make copies of themselves), and "nanocomputers" (basic computational devices constructed out of molecules and atoms).[8]

Drexler also outlines the basis for a medical application of nanotechnology, what nanotech researcher Robert Freitas has termed "nanomedicine."[9] As a specialized field within nanotechnology, nanomedicine works toward bodily repair and even enhancement through the use of engineered, in vivo probes and sensors that would operate, in a semipermanent fashion, within the body. Freitas published the first technical paper in nanomedicine for the "respirocyte," or artificial red blood cells that are capable of advanced filtering and diagnostics (a research project we will return to later).[10]

Since the early 1990s, nanotechnology research has produced some significant empirical results, many of them within the domain of nanomedicine.[11] This is partly because of the development of new instruments for positioning and visualizing atoms, and also because of an influx of research funding. For instance, in 1990 researchers at IBM succeeded in spelling out the company's name by positioning thirty-five xenon atoms; in the early 1990s, researchers at Japan's NEC constructed nano-scale wires out of carbon atoms, which were shown to successfully conduct electricity; and, in late 2000, a team at Bell Labs in New Jersey was able to construct a molecular "motor" out of DNA.[12]

As Drexler and other proponents are aware, the implications of nanotechnology are far-reaching. If it is possible to obtain precision control over matter at the atomic level, then it is possible, quite literally, to make almost anything. Drexler has prophesized a coming era of global abundance, affecting manufacturing, information technologies, health and medicine, the environment, and defense. Such grandiose claims

Figure 14. Ring of iron atoms arranged on a copper surface imaged using Atomic Force Microscopy. Reproduced courtesy of IBM Research, Almaden Research Center. Unauthorized use not permitted.

were undoubtedly part of what sparked the government's interest as well. In 2000, President Bill Clinton publicly endorsed a $500-million National Nanotechnology Initiative (NNI); the Initiative's report is subtitled "Leading the Way to the Next Industrial Revolution."[13]

From Proteins to Nanomachines

A curious thing to note, however, is that, despite nanotechnology's pervasive rhetoric of industrialism and miniature machines, the actual inspiration for the field comes not from engineering, electronics, or materials technologies, but from molecular biology.

In a 1981 technical article presented to the National Academy of Science, Drexler shows how protein production in the cell provides a jumping-off point for the development of more sophisticated means of controlling matter at the atomic level.[14] For Drexler, protein production in the living cell provides what is referred to as the "proof of concept" of nanotechnology: that the ribosome in the cell provides an example from nature that molecular-scale control and engineering is indeed possible. As he states:

Biochemical systems exhibit a "microtechnology" quite different from ours: they are not built down from the macroscopic level but up from the atomic. Biochemical microtechnology provides a beachhead at the molecular level from which to develop new molecular systems by providing a variety of "tools" and "devices" to use and to copy.[15]

This view is key for nanotechnology, because it contextualizes an emerging technology and technical outlook within a decidedly biological framework. Although this folding of tropes upon each other is not new, what Drexler does is to extend the tropes into the domain of engineering practice. The ribosome, for instance, has for some time been known as the "factory" of the cell, taking in RNA and turning out proteins. Only when Drexler refigures this as a technical issue does nanotechnology begin to take shape as an interdisciplinary engineering field. The implication of Drexler's point of view here is that "nature," and the biological domain itself, contains an implicit, inherent engineering principle. In this way, any given biomolecule or biochemical process can be viewed from the perspective of mechanical and electrical engineering, as so many sensors, effectors, rotors, pulleys, rods, circuits, and gates, which collectively perform the ongoing functions in the cell. From this perspective, all of biology is arguably a form of nanotechnology.

Yet, what begins as an inquiry into the molecular systems of protein production ends up as a program for a miniaturized industrial factory. Drexler's vision of nanotechnology some ten years later in his book *Unbounding the Future* is markedly different:

Nanotechnology will be a technology of new molecular machines, of gears and shafts and bearings that move and work with parts shaped in accord with the wave equations at the foundations of natural law.[16]

For Drexler, the process of protein production in the cell serves as a template for research into a "molecular engineering technology." As he points out, such a technology would not rely on the large-scale construction of very small nanomachines, but would rather work in a "bottom-up" fashion by focusing on the precise control of matter at the atomic level. This bottom-up approach not only requires an understanding of molecular interactions, but it also implies the development of novel, nano-scale tools for the construction and manipulation of individual atoms. Nanotechnology would thus encompass both its products and its modes of production.

In this research, which begins by discussing protein production in the cell, Drexler eventually moves toward a machine view of the analogies between microbiological components and industrial technologies: thus RNA acts as a kind of conveyor belt, enzymes acts as clamps, ribosomes act as construction sites, microtubules as struts and beams, and DNA as a storage device. This is not exclusively a linguistic shift; it is a technical shift as well, affecting the kinds of research practices that are developed.

This emphasis on nanotechnology as a new type of industrialism is not only a shift from the biological (protein production in the cell) to the technological (the construction of miniature factories). It is also a shift from living to nonliving matter. Nanotechnology researchers are explicit about their interests in the instrumental character of being able to structure matter at the atomic level. As Drexler states, "improved molecular machinery should no more surprise us than alloy steel being ten times stronger than bone, or copper wires transmitting signals a million times faster than nerves."[17] To paraphrase Drexler, molecules matter, because matter is made of molecules. From the perspective of nanotechnology, all the objects that compose our physical world are specific arrangements of matter in bulk quantities. Arranged one way, matter is coal; arranged another way, it is diamond. The nanotechnical approach is thus to work at the atomic level, constructing and arranging atoms into particular configurations in the attainment of certain goals. These goals can be engineering-based (production), they can be medically based (nanomedicine), or they can be used for a range of other purposes (environment, surveillance, "utility fog," and so on). For nanotechnology, all objects—including the body—are simply arrangements of atoms. It then follows that making any object requires a technology with a precision control over atomic arrangement—the capability to structure matter atom by atom.

In this sense, nanotechnology is indeed a new type of industrialism: it adopts the conventional position of developing techniques for working upon the natural world, including the human body. Nanotechnology is not so much interested in being able to artificially produce proteins, but it is invested in the production of protein-designing machines.

Programmable Matter

What makes such a position possible for nanotechnology is a view of the material world that resembles a kind of atomistic reductionism. This is perhaps the most evident in nanomedicine. As Freitas defines it, "nanomedicine may be broadly defined as the comprehensive monitoring, control, construction, repair, defense, and improvement of all human biological systems, working from the molecular level, using engineered nanodevices and nanostructures."[18] Research in nanomedicine—both theoretical and empirical—has explored the development of in vivo biosensors, "respirocytes" or blood probes, DNA motors and actuators, carbon nanotubes, and tools for nanosurgery. In a more visionary vein, Freitas suggests that the future of nanomedicine may make possible medical interventions into biomolecular regeneration, augmented immune systems and metabolisms, and even life extension and intelligence augmentation:

> nanomedicine will give us unprecedented systemic multilevel access to our internal physical and mental states, including real-time operating parameters of our own organs, tissues, and cells, and, if desired, the activities of small groups of neurons.[19]

Such practices—both the future visions and the current empirical research—not only reconfigure biological matter as amenable to engineering, but they are also predicated on certain notions of health and disease. Speaking about the medical benefits of nanotechnology, Drexler puts it plainly:

> The ill, the old, and the injured all suffer from misarranged patterns of atoms, whether misarranged by invading viruses, passing time, or swerving cars. Devices able to rearrange atoms will be able to set them right.[20]

In this vision, disease is assumed to be essentially an external threat, and, for Freitas, medicine becomes a linear narrative of fighting a battle for health. To this notion of disease, Freitas adds another, what he calls the "volitional normative" model. The volitional normative model combines a view to health as something that technology can conquer, along with a kind of patient-centered, consumer-choice perspective on treatment. Health is viewed as an ideal to be attained through technical means. Its goal is "optimal operation" through "biological programs." Disease thus has molecular roots in undesirable arrangements of atoms. A cure is thus engineering, or better, design. Patients become active in the process by first having at their disposal an array of (design) choices, and then by being able to custom-tailor their own treatment.[21]

What is the body in nanomedicine? The body is a particularized "arrangement of atoms" that has the macroscale effect of being designated as healthy, normal, or diseased. The body is thus like any object—a complex aggregate of atoms that participates in the universalized composition of matter. In such an instance, bodily disease becomes an error in molecular construction, and the nanomedical use of in vivo probes and biosensors becomes a three-dimensional pattern-recognition system. Writing about nanomedicine, Drexler summarizes by outlining three main areas in which nanotechnology can affect the body: as a "workyard" for regulation (maintenance of cellular metabolism), as a "construction site" for regeneration (nano-enhanced molecular and cellular regeneration), and as a "battleground" for self-protection (boosted immune systems, nano-designed blood cells).[22] These three types of nanobodies—the bureaucratic, industrial, and military models—are inflected in biological processes of metabolism, regeneration, and immunity. Freitas broadly reiterates this view:

> nanomedicine phenomenologically regards the human body as an intricately structured machine with trillions of complex, interacting parts, with each part subject to individual scrutiny, repair, and possibly replacement by artificial technological means.[23]

Nanomedicine—and nanotechnology generally—is thus predicated on a view of the body that is open to the interventions of medical design and engineering at the molecular level. In this approach, the "body" is understood as "matter" on two coexistent levels. First, the body is taken as natural matter, as the biological, living body that interacts constantly with nonbiological, nonliving objects—this would constitute the body as the object for nanomedical intervention. But, the body is also taken as technical matter, in a reductionist move that states that, at bottom, both living and nonliving, biological and nonbiological entities are essentially matter. This latter view opens the material world—and the body—to the affordances of instrumentality at the atomic-molecular level. Again, Drexler's use of tropes of industrialism is key: just as modern technology can be seen as a macroscale manipulation of atoms (the natural world), so nanotechnology can be seen as a more refined, more articulate "arrangement of atoms" for a variety of uses. This view of engineering is highly mathematical and combinatorial in its origins; from the nanotechnologist's point of view, technology becomes a means of combinatorics, a means of rearranging matter for particular ends. The body becomes what we might call "programmable matter," a materiality characterized by a constructionist logic, and a highly discrete, combinatoric mutability induced through the intersection of molecular biology and mechanical engineering.

The notion of programmable matter is an intersection between several disciplines, and their relation to the biological domain. From the engineering sciences, programmable matter means a technologically enabled control over the use of matter as resource (and thus the view of nanotechnology as a type of industrialism). From the biological sciences, programmable matter points to the empowering knowledge gained from an understanding of the intricate workings of biomolecules in the body; the "molecular factory" of the ribosome, RNA, and protein chains become a model for nanotechnology, as well as a site of intervention for nanomedicine. Finally, from computer science, the trope itself of matter as "programmable" implies not just instrumentality, but a back-end–front-end functionality, in which access to a "source code" and an understanding of a requisite programming language are what is needed for nanotech to achieve its goals. In other words, nanotechnology, as Drexler, B. C. Crandall, Ralph Merkle, and others argue, is not so much about the direct manipulation of matter at the atomic and molecular levels. Instead, in its development of techniques for rendering matter "programmable," it aims to develop nano-scale systems that themselves work on matter at the atomic and molecular levels.[24]

Bio/Technical

If an atomistic reductionism makes possible a view of the material world as "programmable," then what does this say about the status of the body and biological materiality? If biological matter, and indeed matter generally, can be designed, when does a body become more than a body? To address this question, we can consider two examples of nanomedical research, which stand in a kind of dialectical relationship to each other.

The first is an example of what is often referred to as "exploratory engineering" or "exploratory design." In nanotech research, this means bringing together knowledges from separate scientific fields into a single theoretical research program.[25] In nanomedicine research, this would involve bringing together research from molecular biology, biochemistry, mechanical engineering, and quantum physics, for instance. Although the research in these separate disciplines is not related, the exploratory engineering approach in nanotech research draws upon it in order to construct a plausible theoretical framework for potential empirical research.

An early and thorough example of this is Robert Freitas's work on nano-engineered red blood cells. In his paper "Respirocytes: A Mechanical Artificial Red Cell," first published in 1996, he adopts an exploratory design approach for an in vivo mechanical artificial red blood cell (what he refers to as the respirocyte). In Freitas's words, the respirocyte is a "bloodborne spherical 1-micron diamondoid 1000-atm pressure vessel with active pumping powered by endogenous serum glucose, able to deliver 236 times more oxygen to the tissues per unit volume than natural red cells."[26] Without elaborating on the details of this short description, we can already see that the respirocyte is constructed from nonbiological materials (principally diamond), that it is "powered" by biological molecules (glucose or sugar), and that it aims to function at a higher level than natural red blood cells (able to deliver greater quantities of oxygen and more efficiently manage oxygen–carbon dioxide levels in the body). The very design of this nanodevice therefore brings together two orders of materiality—the biological and the mechanical—in complex ways (Figure 15).

Although there has been medical research on the use of blood substitutes from a biological standpoint, Freitas's work on the respirocyte approaches red blood cells as devices for transporting substances such as oxygen within the bloodstream. (Hemoglobin, the primary protein in red blood cells, interacts with oxygen and carbon dioxide, transporting these gas molecules between the lungs and other tissues of the body.) This functionalist approach frames Freitas's design, so that red blood cells are viewed as gas transport mechanisms. Building upon the extensive research into the biochemistry of gas transport in red blood cells, Freitas notes that there exists a reciprocal relationship between hemoglobin's affinity for oxygen and carbon dioxide. When oxygen pressure levels are high, hemoglobin binds with it to form oxyhemoglobin, releasing carbon dioxide; conversely, when carbon dioxide levels are high, hemoglobin binds with it to form carbamino hemoglobin, releasing oxygen.[27] For Freitas, this reciprocal relationship provides the basis for the design of a sensor-based system that would dynamically interact with these two fundamental molecules within the body. These two conceptual steps are key points in Freitas's design. The first step enframes a biological entity (red blood cells) in the terms of mechanical engineering (transport device). The second step, extending from this, is to articulate the logic driving the device, which similarly involves enframing a biochemical relationship in terms of engineering (hemoglobin as a cybernetic mechanism, operating through sensors, effectors, and feedback). Such initial conceptual steps therefore form a specific relationship between biological and mechanical materialities (Figure 16).

As a gas transport device, which functions through an input-output (sensor-effector)–based mechanism, the "natural" red blood cell therefore provides an impetus for nano-design. Again, Freitas's design is based on a spherical structure that dynamically exchanges oxygen and carbon dioxide molecules with its milieu. As his diagrams show, Freitas's design includes a diamondoid enclosure, a "ballast chamber" (for buoyancy control), gas storage chambers (for oxygen and carbon dioxide), a glucose-drive motor, and even a simple onboard "computer" (for processing gas pressure level changes). However, from the standpoint of programmable matter, the most interesting com-

Figure 15. Proposed external design for respirocytes.

ponent of the respirocyte design is the "molecular sorting assembly," the key to the respirocyte's input-output configuration (Figure 17).

As the figure shows, the molecular sorting assembly is the actual mechanism that exchanges gas molecules between the respirocyte device and the bloodstream. It is composed of a set of molecular "rotors," rods whose ends contain molecular binding sites. These binding sites function just like binding sites at the surface of cells. At the surface of cells are specialized proteins (often called "receptors"), whose specific shape is tailored to recognize only certain types of molecules (e.g., a particular class of enzyme for processing glucose inside the cell). Any molecule that does not exactly "fit" the receptor's shape (like a lock and key) will not be able to permeate the cell membrane and enter the cell. This preciseness of fit is common to all proteins, and is generally referred to as its "specificity." When the precise type of molecule does come into contact with the receptor, the receptor initiates a series of molecular-structural changes (often called a "signal transduction cascade"), enabling the molecule to be brought into the cell, across the cell wall. By attaching selected oxygen and carbon dioxide receptors to the ends of the rotor rods, Freitas is proposing the creation of novel binding sites that are based on his nano-mechanical approach to the red blood cell.[28] In this design, the binding sites act as "sticky ends," only attaching to either oxygen or carbon dioxide molecules.

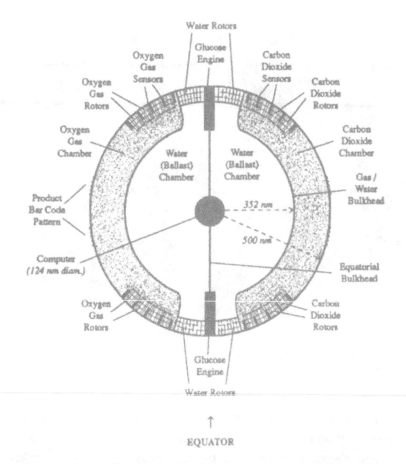

Figure 16. Cutaway diagram of respirocyte, showing nano-constructed chambers and sorting devices.

The rotor binding sites are complemented by molecular sensor rotors. Based on the same design principles as the rotors in the binding sites, these perform basic loading and unloading functions, as well as regulating total gas pressure in the respirocyte's chambers. They regulate the quantitative amount of gas molecules in an intermediary gas chamber, which acts as a kind of "sampler" of the gas concentration in the milieu. As the respirocyte moves through the bloodstream, it tests the milieu by bringing gas molecules into the sampling chamber through the sensor rotors. Once inside the sampling chamber, the sorting rotors exchange gas molecules according to the reciprocal hemoglobin relationship described earlier. The equalized concentration of oxygen versus carbon dioxide molecules is then released in the bloodstream via the effector rotors.

Although Freitas's design has been made specific to the type of nanodevice he is proposing, this basic approach can be seen in a number of similar nanotechnology research projects. The initial approach proceeds by way of an atomistic reductionism;

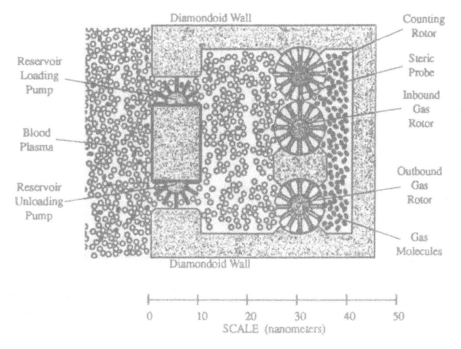

Figure 17. Detail of respirocyte molecular sorting assembly. Copyright 1996 Robert A. Freitas Jr. All rights reserved.

in fact, this is less an approach and more an assumption in nanotech research. As we have seen in Drexler's formulation, in nanotechnology the body is viewed as a particular "arrangement of atoms." This assumption is largely what enables the view of matter as programmable. Not only can nanodevices be constructed atom by atom, but, more important, the way in which they are engineered can specify their functionalities and the range of what they can do. "Programmable matter" is therefore to be distinguished from a mere constructionism, the basic task of building something whose exact functions are known and can be predicted.

By contrast, Freitas's respirocyte, and similar proposed nanomedical devices, have as their aim the design of nanosystems that operate within and as part of the biological body. The respirocyte is not a one-symptom, single-injection therapy; that is, arguably, the approach of many current biomedical techniques. Rather, the respirocyte is designed to function on a long-term basis as part of the body's system of red blood cells. The design of the molecular sorting assembly, with its sensor and sorting rotors, does not have as its goal the treatment of a symptom, but rather the ongoing regulation of particular biological processes (e.g., oxygen transport in the blood). The key to understanding nanomedicine's approach of programmable matter is to understand this distinction. Freitas does outline possible medical uses for respirocytes: blood transfusions, treatment of anemia, fetal asphyxia or related disorders, respiratory disease, as well as medical diagnostics for tumor detection.[29] However, the projected use of respirocytes here is regulatory rather than symptomatic; the respirocytes function as long-term augmentations to the natural-biological red blood cells in the body.

This approach, in which programmable matter is understood as biological regulation (and not simply the treatment and targeting of symptoms), raises two difficult issues for nanomedicine, and nanotech generally. The first issue has to do with the biological implications of a conceptual mixing of "body" and "machine," or biological and nonbiological components. Clearly, the very introduction of designed, "artificial" red blood cells is in some sense an augmentation of the biological body's regulatory systems. This is as much borne out by the function of the respirocytes (that they are more efficient) as it is by their very construction (from "nonbiological" substances such as diamond).[30] Indeed, one of the primary challenges that Freitas expresses is the issue of biocompatibility. Not only do substances such as diamond require possible modification in order that they not immediately be (mis)recognized as "foreign" by the immune system, but it is uncertain how the immune system would respond to an "artificial" change in the body's dynamically regulated count of red blood cells. Specifically, how the introduction of respirocytes would affect the production of red blood cells (or "erythropoiesis") is largely unknown. This, of course, requires a consideration of the body at the molecular and nano levels as a system, because a number of red blood cells die and are replaced daily, and erythropoiesis itself responds to signals from a variety of sites in the body (e.g., the kidneys, the heart/circulatory system).

The second issue has to do with the distinction between medical therapy and biological augmentation. Alongside the medical applications of the respirocytes, Freitas also mentions their use for in vivo medical imaging, respiratory augmentation for underwater breathing, and even as an aid for aerobic activity (such as physical performance in sports).[31] Like many statements about nanotech, such proposed uses are intended to span the range of possible application, from the very practical to the more "visionary." In this case, the practical-medical is clearly demarcated from the visionary—not by its feasibility, but rather by its proximity to clearly articulated medical application (where the "medical" stands in for the "practical"). Here ethical issues come to the fore, in terms of, in an emerging field, what counts as a valid technoscientific intervention, and what counts as either vanity, unnecessary modification, or an intrusion of the technical on the natural. In the way Freitas has presented his work, it is clear that medical uses (such as in the treatment of anemia) belong to valid application, whereas the use for breaking world records in sports belongs to an extramedical (or excessive) application.

Without belaboring ethical slippery slopes, we can note that this tension also resides within the respirocyte design itself. As we have seen, the respirocyte is a complex of biological and nonbiological substances and processes. Being constructed from diamondoid materials, and housing nano-scale gears, rotors, and engines, the respirocyte little resembles actual red blood cells. Similarly, in the use of multiple gas sample and storage chambers, sorting and sensor rotor systems, the respirocyte aims not merely to mime the red blood cell, but in fact to surpass it in its functioning (greater quantity of delivery, greater efficiency, ability to externally control respirocyte activation). The respirocyte presents us with a strange intersection of the biological and the technological: on the one hand, it is clearly a nonbiological, "artificial-mechanical" device. On the other hand, its very purpose is to integrate itself seamlessly into the bloodstream, thereby functioning alongside biological red blood cells.

On the level of both substance and process, form and function, the respirocyte design can be seen as an attempt by nanotechnology to literally "make sense" of the division between living and nonliving matter. The respirocyte, as Freitas states (echoing Drexler), takes its inspiration from the biological model of the red blood cell. Freitas asks, from the engineer's vantage point, "What does a red blood cell do?" The red blood cell functions as a gas transport device. The design of the respirocyte moves on from this point, by combining biological and nonbiological components into assemblies such as the molecular sorting rotor (combined of rotor rods and biomolecular receptors, to form a nanotechnical "binding site"). From the nanotechnological perspective ("all matter is matter"), the combination of functionalities from mechanical engineering (rotors) and biochemistry (receptor proteins) attempts to have the best of both worlds: a compatible interface between device and milieu, as well as a seamless internal functionality. At once external to the biological body and integrated into its functioning, the respirocyte of a nanotechnological approach takes the biological as its model, while providing a technological solution, whose aim is to function biologically. In short, we can say that the respriocyte provides us with an example of one strategy in nanotechnology: the biological-as-model, the technological-as-solution.

From this point, we can briefly examine a second example that provides a rough analogue to the example of the respirocyte. While the respirocyte provides a technological (mechanical engineering) response to a biological model (red blood cell), the example of "DNA motors" offers a biological response to a technological model. Researchers, working separately at Cornell University, Oxford University, and Bell Labs in New Jersey, have each empirically demonstrated the feasibility of nano-scale "motors" constructed out of DNA and other biological molecules.[32] Such research is contributing to the larger goal—fully equipped computers assembled at the nano-scale. The development of DNA motors or actuators is one research area in this direction, and other areas of research are, in a like manner, exploring the "nonbiological" uses of DNA, proteins, and other biomolecules as storage devices, transistors, and actuators.

Taking the Bell Labs research on DNA motors as an example, we can see how another strategy in nanotechnology research operates almost opposite to the prior example of artificial-mechanical respirocytes. In 2000, a research team at Bell Labs headed by Bernard Yurke published an article in *Nature* describing their experiment on a DNA-powered motor. Their DNA motor is composed entirely of four DNA single strands.

Figure 18. DNA motor, showing DNA hinge, binding of fuel DNA, and full binding of both DNA strands. Reproduced courtesy of Lucent Technologies Inc./Bell Labs.

The first strand is a short (~ 30 base pairs) "backbone" strand, to which are attached two more DNA "peripheral" strands. The two peripheral strands are separated by four open-ended base pairs on the backbone strand, enabling the complex to form a "V" shape (see Figure 18). This middle section of unbonded base pairs forms a "hinge" for the motor, and the ends of the backbone strand are "sticky ends" of base pairs that are used to attach the fourth strand, the "fuel." The fuel strand of DNA, when it comes into contact with the backbone-peripheral strand complex, immediately bonds to it, according to the classical rules of DNA base pair complementarity elaborated by James Watson and Francis Crick in the 1950s.[33] When this happens, the DNA motor's ends close like tweezers, thanks to the bonding of the fourth, fuel strand of DNA. This process of annealing (attaching) and denaturing (detaching) DNA can be repeated using standard molecular biology laboratory protocols (heating, adding reagent, cooling).

The Bell Labs team envisions several uses for their DNA motor, including information processing (using base pair complementarity as a kind of "gate"), as a scaffolding for computer circuits at the nano-scale, and, most prominently, as tiny tweezers for the DNA-driven assembly of nano-scale circuits. This last use resonates with Drexler's proposal for nano-scale "assemblers" in his *Engines of Creation*.[34] But although these uses have yet to be demonstrated in the lab, the results from this and other similar experiments suggest another approach in nanotechnology, one in which nanodevice design, assembly, and implementation derive solely from biological materials themselves. Like Freitas's respirocyte design, the DNA motor also takes its inspiration from "nature," in particular from DNA (as opposed to red blood cells). Yet, in contradistinction to the respirocyte, the DNA motor does not attempt to redesign, improve, or augment the biological functioning of DNA in the cell (as the respirocyte does for the red blood cell). Rather, the DNA motor takes a specifically technical inspiration from DNA: the ways in which the biochemistry of DNA base-pair complementarity can be reutilized toward the construction of a nano-scale assembler component.

As Yurke, the Bell Labs team leader, states, "We're using DNA in a very non-biological way—as a structural material and a fuel in a sense."[35] This approach, like Freitas's approach with the respirocyte, also asks, "What does it do?" However, because Freitas's goal is "biological"—that is, the design of biologically functional artificial-mechanical red blood cells—this question is contextualized within a biological/cytological framework. By contrast, the Bell Labs' asking of the same question is only biological inasmuch as the biological is seen through a technological lens. The Bell Labs aim is not biological, nor medical, but rather technological (and more specifically, in the long run, computational). Whereas the respirocyte takes the biological-as-model, and responds with the technological as a solution, the DNA motor takes the technological as its model (actuators, transistors), and responds with a solution that is biological and biochemical (DNA base pair binding).

Moreover, as with Freitas's work on the respirocyte, there also exists an internal tension with the DNA motor. This tension primarily revolves around the issue of context. DNA base pair complementarity has had a wide range of "non–natural" uses, from genetic engineering's use of bacterial plasmids to insert foreign DNA, to its use in DNA computing experiments, to the examples of nanotechnological DNA motors described here. In each case, the assumption is that this "natural" phenomenon, which occurs regularly in the DNA of living cells, is "portable" (in computer software terms) to other contexts. However, as molecular biologists have known for some time, the single biochemical phenomenon of DNA base pair binding hardly ever takes place in isolation; it is always connected to a myriad of other processes, playing a role in cellular metabolism, replication, and protein synthesis. This is not to argue that DNA base pair binding cannot be abstracted technically: the research into DNA motors (as well as DNA computing) shows that it can. Rather, the question is what happens to the categories of the "biological" and "technological" in examples such as the DNA motor, where "matter" constantly crosses boundaries between the two.

The question of how the "biological" is to be framed, and how it is to be related to the "technological," is raised here along several lines. In one sense, the DNA motor is fully biological, if by this we place an emphasis on the components present in a given system. In another sense, the DNA motor is fully technological, if by this we mean a system's function rather than its components. In still another way, the DNA motor is thoroughly biotechnical, in that, like biotechnology generally, it makes novel uses of biological processes such as DNA base pair binding and PCR (for automating the replication of selected DNA strands). Finally, the DNA motor is broadly nanotechnical, through its specific engineering of technological systems at the molecular and atomic levels.

We might summarize this by saying that nanodevices such as the DNA motor place "biology" and "technology" in manifold relationships, in which one is constantly folded onto the other. At times, the criterion for establishing such relationships is substance or the components; other times it is processes or function. In each instance, what counts as "biological" is largely dependent on the way in which "matter" is contextualized by nanotechnology's engineering viewpoint. As we have seen with the issues of biocompatibility in the respirocyte, and the technical repurposing of DNA in the DNA motor, both suggest that the notion of what counts as "biological" or "technological" is constantly transformed in fields such as nanomedicine and nanotechnology.

Programmable Matter 2.0

But is there not something highly contradictory about nanomedicine's view of matter in these instances of the respirocyte and the DNA motor? In other words, in the move away from living to nonliving matter, from proteins to nanomachines, does not nano-medicine undermine its claim for the universality of matter?

It is in this relationship between living and nonliving matter that nanotechnology becomes especially interesting. From one perspective, it is clear that Drexler departs from the biologistic model of the ribosome in his vision of nanotechnology. In this case, nanomachines play the role technology traditionally has: as an instrumental tool for working upon the natural world. Here, technology operates in an autonomous manner upon its objects, creating new relationships and new formations from their raw material.

In another sense, however, when we work at the nano-scale, the accepted differ-ences between living organisms and inert objects becomes, to some extent, irrelevant. What matters is not the macroscale status of an object (human, chair, computer), but rather the types of molecular characteristics that such objects express at the nano-scale. The basic differences between solids, liquids, and gases is a prime example here (ice, water, air).

What Drexler appears to do is to universalize matter as a means of establishing a privileged site for a technology of the control over matter. What is not included in Drexler's formulation, however, is any degree of reflexivity, in which the "matter" of the technologies themselves (the nanomachines that do the constructing from the bottom up) is explicitly part of this general characterization of matter-as-matter.

The challenge put to nanomedicine is this: In order for a nanomachine such as an in vivo respirocyte to operate diagnostically, researchers must have a sense of a clear difference between the nanoprobe and the blood cells whose concentration it measures. In other words, for nanotechnology to operate instrumentally, a distance must be established between technology and its object. Yet, because nanomedicine involves the use of engineered nanomachines operating inside the living body, possibly in a per-manent or semipermanent way, the nanomachines must be able to successfully inte-grate themselves into the body's molecular surround.

The result is that nanomedicine requires the incorporation of medical technologies into the biological body-as-self, so that those technologies can operate instrumentally as not-self. This is a dual strategy of being able to precisely control matter, and also being able to absorb its benefits in vivo. Drexler's original move from proteins to nanomachines, from the biological to the technological, seems to present us with a contradiction, or at least a tension, in how matter is defined in the medical application of nanotechnology. Either matter is a universal, effacing the macroscale distinctions between living and nonliving matter, or matter is highly differentiated and heterogeneous, in which there may even be incommensurabilities between living and nonliving matter. While the former view provides us with a world that is constructed from a universal set of components, the latter view provides a distancing effect with which to approach that world instrumentally.

This tension within nanotechnology and nanomedicine leads to a reconsideration of the ways in which matter is endowed with "life." Whether it is a connection based on substance, structure, function, or some other quality, the relation between matter and life is one of the long-standing, contentious issues in the biosciences. By performing an atomistic reductionism, nanotech conceives of the body as a particular type of programmable matter, one open to intervention in regulation, regeneration, and self-protection. However, this link between matter and life remains unarticulated in nanotechnology discourse, and it is to this relationship that we now turn.

Postvital, Postnatural

Recent discourses that explore the interrelationships between human subjects and technologies are filled with different "post-" words, from postmodernism's emphasis on technologies of spectacle and simulation to posthumanism's exploration of the ways in which research in fields such as artificial life and intelligent computers is transforming our notions of the ontological separation between humans and machines.[36] Although there is certainly a sense in which the proliferation of "post-" terminologies can have a deadening or predominantly surface-level effect, such a discourse also points to the early debates on postmodernism's ambivalent meaning of *post* as either coming after, departing from, or continuing and intensifying.

Along these lines, we can briefly bring in two more "post-" terms, the "postvital" and the "postnatural," as ways of understanding the specific modes whereby nanotechnology attempts to manage the relationship between biologies and technologies. Nanotech is unique because, although it can be considered a postvitalist and a postnatural science, it also, inadvertently, frustrates the tendency to view it as the next progressive stage in the advancement and improvement of a future technologized approach to matter-as-engineered. Ultimately, the postvital and postnatural aspects of nanotech and nanomedicine signify a more general ambivalence between a privileged definition of "life" and a wide knowledge base through which that same "life" can be reduced to the level of programmable matter.

In an analysis of the rhetorical modes of molecular biology, Richard Doyle shows how developments within modern biology transform its object, the study of "life." If histories of the so-called life sciences often mention the shift from natural history's taxonomies of living beings (based on the systematic comparison of visible characteristics) to modern biology's study of something essential of the organism called "life," Doyle points to the tensions within modern biology between a life that can be scientifically explained and a life that remains an inaccessible, central part of the organism.

The discontinuity Doyle seeks to highlight is the relationship between a vitalist biological science—which claims an essentially unknowable core of the organism, its "life force"—and the gradual shift toward molecular biology, whose primary claims have to do with the interpretation of life on the molecular level. As Doyle states:

> Despite their apparent opposition, both vitalism, the idea that life exceeds known physiochemical laws, and molecular biology, the science that has claimed the reductions of life to those same physiochemical laws, relied on an unseen unity that traversed all the differences and discontinuities of living being, "life."[37]

Thus, in the debates between nineteenth-century vitalism and an emerging science of biology, a commonality of "life" "becomes the unseen guarantor of biology, knowable only at a distance."[38] Doyle's main point here concerning molecular biology—which has come to mean the study of DNA in the organism—is that even as it reduces the organism to a set of molecular codes or DNA sequence, it continues to assume that its object of study is something more than DNA sequences itself; at the same time that molecular biology aims to study DNA as its "central dogma," it also reserves something in excess of DNA itself as the core of organismic life.

The studies of Doyle, as well as of Lily Kay and Evelyn Fox Keller, show how molecular biology forces the molecule of DNA to carry a meaning far beyond the study of the molecule itself; DNA becomes the "book of life," the "code of codes," the key to understanding not only the organism's meaning as living, but also to the organism's future, the DNA that codes for visible characteristics not yet visible.[39] The contradiction at work here is that while molecular biology appears to adopt a rigorous empirical approach to the organism on the molecular level, showing that an organism can be understood through a study of its DNA, it also utilizes this more precise, increasingly informatic approach to query the larger mysteries of the organism. The notion of a genetic "code" thus serves to reduce the organism—vitalism's life force—to sequence and code, while also adopting this approach to bring "life" to the surface level of linearized genetic sequence.

It is this doubled action within the discourse of molecular biology—the reduction to code, and the coding of life—that produces a unique type of organism in molecular biology, which Doyle refers to as the "postvital" organism. Thus molecular biology's notion of the organism as DNA "becomes possible with the arrival of the post-vital body, a body in which the distinct, modern categories of surface and depth, being and living, implode into a new density of coding, what Jacob [François Jacob, codiscoverer of messenger RNA] calls the 'algorithms of the living world.'"[40]

The postvital body is, for modern life sciences, both a technical and a philosophical feat. It is a technical feat because it appears to dispel the romanticism of nineteenth-century vitalist theories of life by demonstrating that the organism can be reduced to (a genetic) code, a particular mathematical, informatic textualization of the organism. But for this same reason it is also a feat within the ongoing concerns of the philosophy of biology, for it shows that this very reduction of the organism to (genetic) code is itself a making apparent, a bringing to the surface, of the "truth" of the organism. As Doyle states, this doubling of suggesting that there is nothing beyond genetic code, and that within the genetic code are the "secrets" of life, provides for a kind of "resolution" to the organism. This is a resolution that intimates the greater resolution of the organism technically (the mapping of the organism as a genome) as well as a resolution to the hunt for the secrets of the organism as a living entity, "a story of resolution told in higher and higher resolution."[41]

For Doyle, the logical extension of the post-vital body lies in the adoption of information tropes by molecular biology, the genome as software, a program of a finite number of elements that produce, through their combinatorial, mathematical functioning, incredible complexity (once again displacing the object of "life" into mathematical complexity). Thus, when researchers in artificial life begin to model open-ended, emergent computer programs along evolutionary and adaptive lines, the very practice of computer science itself looks back to the vitalist concern with an ungraspable element of "life"—unpredictable, emergent, and inaccessible to systematization.

This intimate coupling between vitalist life force and DNA code—where the latter both denounces and reinterprets the former—brings up the degree to which sciences such as molecular biology can be said to be studying "nature," when their central tropes are borrowed from information theory and cybernetics. In an essay exploring the scientific implications of a "digital aesthetics," Sean Cubitt proposes the term *postnatural* to describe technoscientific approaches to "life" that claim to surpass, enhance, or move beyond the organic.[42] Cubitt's distinctions between the supernatural (concerns with life after death, ghosts, specters), the antinatural (instrumental uses of technology over and against the natural world), and the cybernatural (the creation of artificial life, intelligent machines) present different modes of imagining some unity called "life" that is manifested outside of the organic, or strictly biological, domain.

But each of these—which remain coexistent discourses—also reconfigures what is meant by "nature" in its imagining of a state beyond nature, outside the organic. Thus the supernatural often implicates itself as latent in the natural, the antinatural constructs parallel, technologized worlds that are the displacement of cultural desires, and the cybernatural reveals a fundamental theological anxiety over the creation of life. Cubitt's discussion of the postnatural is in part a discussion on the way in which technologies respond to, react against, and reconfigure "nature." In the antinatural discourses of virtual reality, and cybernatural discourses of intelligent machines, is embedded the supernatural fascination and desire for noncorporeal, disembodied life, a life removed from its biologic foundations but still life, now reinterpreted as information, "vital data."

From one perspective, nanotechnology fields that combine genomics, proteomics, and bioinformatics can certainly be characterized as postvital and postnatural. Molecular nanomedicine, itself an outgrowth of postwar molecular biology and biochemistry, is postvital in its preoccupation with genes of value, genes that not only contribute to diseases, but that generally code for the structure and functioning of the body. But such investigations also lead to the extremes of a molecular genetic vitalism, as in the exaggerated future visions of intelligent in vivo nanodevices or immunity-boosting nanoprobes.[43]

Likewise, nanotech is perhaps the exemplary postnatural science, in that it is concerned above all with seeking states of life beyond the natural, biological body. The research into implantable nanodevices, DNA motors, and molecular circuitry are all questions posed to the limitations of the natural body recontextualized as more than natural. In this process, this adoption of a technological view of life, the natural organism becomes more defined by its "natural/biological" limitations—susceptibility to disease and infection, cycles of growth and decay, mortality itself—than by any romantic, vitalist core. Instead of simply displacing the biological, natural body with a totally mechanical or computerized hardware body (the Extropian's dream of "uploading" mind), nanotech promotes an enhancement of the body on the molecular-biological level, in effect improving the biological, extending its capabilities, redesigning the biological to surpass itself.

Thus the bodies of nanotech and nanomedicine—and again it is important to stress that these are defined as an "arrangement of atoms"—are both postvital (resurrecting vitalist concerns via atomistic reduction) and postnatural (attempting to surpass the biological through a technical approach to the biological). Yet there is also something else at work here, a return, similar to the one that Doyle speaks of in the postvital, a return to the vitalist reserve of a life force beyond human knowledge, which serves as both the foundation and the origin of the study of life. If the bodies of nanotech are programmable matter, then we are talking about a particular approach to life, one that takes an informatic worldview as its starting point, a posthuman view that the world can be interpreted as code, can be reduced to code—as Donna Haraway states, "a translation of the world into a problem of coding."[44] This posthuman view is not just an abstraction of the physical world, a liquidation of materiality, but, as Katherine Hayles points out, an equivalence made between materiality and informatics, flesh and data, bodies and information.[45] The posthuman does not so much deny materiality as it chooses to reconfigure the material world itself as inherently amenable to a science and technology of informatics.

This positing of an "equals" sign between elements traditionally at odds with each other ("words and things") results in some very basic ambiguities in the relationship between the human and posthuman subjects in the context of nanotechnology. On the one hand, the posthuman desires the transformative capacities of technology, the improvement of the human through technology. But, on the other hand, the posthuman reserves the right to preserve something called the human as constant and constitutive throughout the process. Here this mode of preserving the human is analogous to the manner in which the concerns of vitalism are maintained in molecular biology, something foundational but also inaccessible that remains unthreatened by radical techno-logical transformation.

Nanotechnology is a unique type of posthumanism because, unlike AI (artificial intelligence) or AL (a-life), its primary concern is not so much "mind" or abstract "pattern," but rather the body viewed at the atomic-molecular level. While adopting an informatic worldview consonant with much posthumanist thinking, nanotech re-search leads not to increased informatic abstraction (uploading software minds), but rather, in a kind of rematerializing loop, back to the body (the body of the very small). But this return to the body is also an extreme mutation of the body, a "super-natural" body, a body more biological than the biological, technically enhanced through the in-strumental use of biological structures and processes. In such cases, DNA acts as more than DNA; it functions technically as a motor, or as a gate for a molecular computer. Gas transport in the blood is not only equilibrated by the homeostatic role of the respirocyte; it also augments the functioning of an entire physiological system in its intended semipermanent functioning.

Although, from one perspective, nanotech and nanomedicine research seems to be an extreme postvitalism (to the degree that the biological can be engineered) and an extreme postnaturalism (to the degree that nature can be improved upon), the way in which it returns to the body—via programmable matter—shows how it is also a re-turn to the dream of being able to make the "truth" of the body show forth, to in effect materialize the truth of the natural, biological body, its very materiality. It is in this sense that biotechnics leads to a reinscription of meaning in the body, "vital data" without which the natural and the biological cannot be understood, without which, in the case of medical application, life cannot survive.

"Here we are, we drift like gas"

Nanotech works on a very, very, very, very small scale—billionths of meters, relations between individual atoms, even quantum dots. When approached from this viewpoint, the concept of production and of materiality itself becomes an issue of molecular engineering. From the nanotechnological perspective, theoretically anything can be constructed, atom by atom, molecule by molecule—Drexler's book contains specula-tive scenarios involving the nanomanufacturing of rockets, human organs, foods, and computer hardware, to name just a few examples.[46] As Drexler states, any object can be seen as a particular arrangement of atoms, including the human body. Under-standing the principles in design and construction will lead to advances in engineer-ing. For nanotech, matter is matter, and precise control at the atomic level can open the way to a new industrial revolution (as the National Nanotechnology Initiative puts it) or a new era of global abundance (as researcher B. C. Crandall puts it).

The approach of atomistic reductionism, and the engineering-based solution pro-posed by nanotech, point to a view of matter as being reconfigured, redesigned, mate-rially edited. In short, nanotech's approach to the body specifically, and matter generally, envisions matter as programmable—matter as not excluding informatics. Programmable matter is a technical approach to the physical world in which the distinction between information and materiality is effaced. For nanotech, the entire apparatus of nanomachines—assemblers, replicators, nanocomputers—is itself built out of the same molecular and atomic elements that compose the physical world. Nanotech's ultimate engineering fantasy—that of the nanocomputer, or a computer hardware apparatus that is assembled from atoms—is a direct example of its will to materialize information.

This confluence of information and matter is an interesting implication, for it works against a long tradition of technical thought that dissociates information from its material substrate.[47] From a certain perspective, nanotech can be seen to be embodying information in molecules; or rather, positing the indissociability of matter from data. Yet, we should point out that this problematizing of the separation between information and materiality is not the primary intention of nanotech research. It comes about almost as an accident in Drexler's thinking, for instance. Nanotech is predicated on the notion that matter is potentially totally controllable through engineering design approaches. In this sense, nanotech is an example of an instrumental technology; an attempt to technically instantiate control over the natural world, or, in this case, the world of natural laws.

On a general level, nanotech research seems to have abandoned the original biomolecular model of protein production in favor of more recognizably machinic systems. In other words, it seems to have incorporated the ideological of instrumental technology into its research and design; thus, instead of amorphous, complex systems such as the ribosome, we have a series of nano-scale machine parts designed to accomplish specific tasks (create a current in fluid surroundings, rotate, clip, puncture, etc.). The goal of first-generation nanotech research is to construct (literally, atom by atom) machines for working on the material world (for instance, nanotubes for the speedy conduction of other molecules). This ideological and technical distancing has pushed nanotech research further from organic design systems, and closer to mechanical ones; in other words, nanotech is reiterating its telos as an instrumental technology by separating itself from nature (nature-as-matter) while also embodying the primal elements of the physical world (and here matter is matter) as instrumental technologies.

Nanomedicine is based on the ontological distinction between the body as natural and the nanosensor as a medical technology. This is despite the fact that the body will incorporate the nanosensors permanently and recognize them as "self," and that the nanosensors may someday be composed of organic as well as nonorganic elements. This division is necessary because of the initial approach in nanotech: an instrumental, artificially constructed apparatus for working upon something pre-given in the world (and thus natural). Without this division, medical practice loses its agency and autonomy, just as the body loses its passivity (to the nanotechnical gaze) and object quality within medical contexts.

But nanomedicine's most powerful potential lies in the possible linkages it may form with biotech research; this is the point at which biotech will become a molecular engineering science, and the point at which nanomedicine will cease functioning as a diagnostic tool, and as a mode for reconfiguring the body itself—biomedia.

We might outline the processes in nanomedicine's confusions surrounding matter in three steps: First, the development of a nanotech field shifts our concerns over technology and nature to one of matter. A technology of matter threatens to efface the boundary between nature and technics, because everything happens with physical laws on the molecular level. Yet, second, this nanotech also reiterates the division between nature and technics because nanotech is a set of tools and techniques for manipulating matter; here matter becomes coded as nature, and assemblers become coded as technology. Third, from here it is a short step to claims for a new industrial revolution.

This boundary confusion on the level of matter is specified within nanomedicine as a relationship between human and machine, mechanomolecules and biomolecules. Although much thinking on nanomedicine tends to reiterate this divide, nanotech provides the opportunity for thinking about the body as biomechanical—not the awkward addition of prosthetics, not the black-box injections of genes or genetic cocktails, but a redefinition of the body as both natural and technical matter.

This possibility implies that, within current thinking on nanomedicine, a series of ideological inclusions and exclusions are taking place. As we have seen in the early visions of nanotechnology (Feynman, Drexler), the principles of engineering have come to dominate thinking in research. Excluded is any nanotechnical system that does not follow the principle of mechanical engineering, despite Drexler's early inspirations from molecular biology. We have also seen that, in this prioritizing of engineering principles, the industrial principles of "molecular manufacturing" have become prevalent in thinking about nanotechnology as (industrial) application. Excluded, therefore, is any nanotechnical process that does not follow the means–ends logic of nanomachines, despite the explicit integration in living systems implied in Frietas's nanomedical designs. Finally, from a biological standpoint, nanotechnology provides an important opportunity for a rethinking of the biological–technological divide, which we can see replicated in disciplinary divisions. Excluded, therefore, is any design process that does not grant autonomy and single-action effectiveness to nanomachines, even though fields such as nanomedicine implicitly work toward a view of the body as both biological and technological. In short, the body as a noncausal, networked, multiagential, and ontogenic system is excluded from the concerns of nanodesign.

In this way, nanotechnology is an example of the privileging of a technological domain through the contradictory regulation of the boundary between living and nonliving matter. Programmable matter is nanotech's approach to the challenge of implementing a total design environment in relation to the physical world. It attends to the challenge of developing novel techniques and technologies, as well as the objects of technical control. Although the initial phases of nanotech research are adopting predominantly engineering approaches, the ultimate goal of nanotech is not directed production per se (bulk machines for constructing nano-scale molecules), but the construction of production systems such as assemblers. Programmable matter is an approach to self-generating systems, a kind of "metadesign" for generating precise configurations of matter.

As proponents of nanotechnology have speculated, the applications for medicine could include everything from cell and tissue repair, to smart drugs, to cryonics and antiaging techniques, opening the way to a posthuman future of biomolecular design. It is in this notion of the body as programmable matter that nanotechnology will come up against some of its greatest challenges. As a still-developing field, the line dividing medical healing and biomolecular design is generally underexamined in nanotechnology. With the promises and the research pointing to a so-called new industrial revolution, nanotechnology will have to confront the tensions between a total control of the material world and the radical transformations that such a relationship between control and design might very well bring about.

CHAPTER SIX

Systems Biology

Parallel CorpoRealities

Centralized, Decentralized, Distributed

In "Swarm," a short story by Bruce Sterling, we witness the intimacies of an alien ethnography. Captain-Doctor Simon Afriel, a member of a future human government alliance called the Ring Council, is sent on an interspecies diplomatic mission to an alien race known as the Swarm. In Sterling's future world, the human race has split into two factions. One, the Mechanists, relies on cybernetics technologies to develop advanced industrial modes of production and half-robot bodies. The other, of which Afriel is a member, are the Shapers, those who have refused cyborg technology in favor of genetic engineering and wetware. In a literal body politics, the two factions are engaged in constant political battles, with their different approaches to the augmentation of the human body determining their different modes of viewing the future of the human race.

Upon entering the Swarm, Afriel is greeted by Dr. Galina Mirny, a scientist who has been living among the Swarm for some time. Mirny gives him a brief tour of a small sector of the Swarm. Filled with innumerable strange creatures, the Swarm operates like an advanced insect colony, with no apparent central control but many points of defined local activity. The creatures range from the human-sized, antlike "symbionts," whose primary function is communication and navigation, to the "warriors," whose primary function is to guard the hatchery, and the Queen, who (despite the human-given title) has no other function than to produce the larval eggs. The Swarm operates in a set of labyrinthine tunnels in a large meteorite. The tunnels are not fixed, but are always changing, new ones being drilled by "tunneling" creatures, old ones being patched up or filled in by other large, furry, disk-shaped creatures.

Afriel's main motive in visiting the Swarm—about which very little is known, except that they are a "primitive" species—is to see whether they possess any resources or novel technologies for the Shapers to make use of and possibly invest in. In a replaying of the historical contexts of European exploration, expansion, and colonization, Afriel goes to live among the Swarm, and their incomprehensibly alien, monstrous, "savage" ways. As Mirny puts it:

> This Nest, these creatures—they're not intelligent, Captain. They can't think, they can't learn. They're innocent, primordially innocent. They have no knowledge of good and evil. They had no knowledge of anything. The last thing they need is to become pawns in a power struggle within some other race, light-years away.[1]

Using chemically synthesized pheromones, Afriel discovers that the various creatures in the Swarm can be easily controlled. Mirny soon learns that Afriel's plan is to take a genetic sample of one of the Swarm eggs, in order to secretly import it back into the Shaper labs. The Shapers' plan becomes clear: import the genetic material for a slave race of creatures without intelligence.

However, amid their experiments, something happens. The Swarm, as a whole, becomes "aware" of the slight imbalance caused by Afriel's experiments with his synthetic pheromones. In a scene filled with imagery of the grotesque, Afriel is forcibly brought before an enormous mass of writhing flesh, distinguished only by a mouth-like aperture, eyes on stalks, and thousands of feelers. It states, "I am the Swarm. That is, I am one of its castes. I am a tool, an adaptation: my speciality is intelligence. I am not often needed."[2] As Afriel learns, this module of the Swarm has been aware of his intentions before his arrival; it has, in fact, confronted this same situation innumerable times. All alien species (alien to the Swarm, that is) that come to the Swarm with instrumental purposes in mind "trigger" the development of the intelligence module, which confronts, and then incorporates, the intruding alien species:

> I am a tool, as I said . . . When you began your pheromonal experiments, the chemical imbalance became apparent to the Queen. It triggered certain genetic patterns within her body, and I was reborn . . . Within three days I was fully self-conscious. Within five days I had deciphered these markings on my body. They are the genetically encoded history of my race . . . within five days and two hours I recognized the problem at hand and knew what to do.[3]

The mass of flesh known as "Swarm" is thus generated by the Swarm itself, through a cascading pathway of genetic signals. Its "intelligence" emerges within days, it sets about resolving a conflict for which it was required, and then, once the conflict is ameliorated, it simply dies, until needed again.[4] Afriel discovers that the fantastic, teratological bestiary of the Swarm is actually a collection of adapted "alien" species that have coevolved in the Swarm over millennia. The human or posthuman specimen of Afriel and Mirny become its latest addition.

"Swarm" presents us with a paradigmatic case of "the other" in the guise of the posthuman. At once repulsive, bewildering, fascinating, and utterly unknowable, the alien bestiary of "Swarm" is a diametrical contrast to the notion of the human, in substance, form, and behavior. Like many such posthumanist science fictions, "Swarm" envisions the posthuman not as a technophilic utopia, but rather as an ambivalent affront to normative assumptions concerning the human in biological, political, technological, and social terms. At its basis, the swarm of activity in "Swarm" confronts the human subject with tensions between a living "system" and the individual body, between "networks" and the self.[5]

Such tensions, dramatized through SF, are also materialized in a different way in contemporary approaches to the study of genomes. Loosely known as "systems biology," such approaches are broadly undertaking research programs dedicated to a more complex understanding of biological life than the current paradigm of gene-centric approaches (e.g., genome mapping, rational drug design, gene therapy).[6] In research carried out by independent organizations such as the Institute for Systems Biology, the Alliance for Cellular Signaling, and the BioPathways Consortium, systems biology is gradually gaining acceptance in mainstream genomics, such that it is not uncommon to see the rhetoric of "systems," "complexity," and "patterns" in biotech advertising.

Based on earlier developments in systems theory and cybernetics, systems biology, while offering alternatives to gene-centric approaches, also brings with it a set of issues that are at once technical and philosophical. Following a consideration of systems theory, and its relation to systems biology, we will take up interdisciplinary research in self-organization and autopoiesis, as examples of mediations between the overgenerality of systems theory and the technological overdetermination of systems biology.

Histories and Systems

The "systems view of the world" appears to have a paradoxical existence. On the one hand, it seems to never have really taken hold, in the end only offering vague metaphors about "holism" and "networks" made all the more abstract by mathematical analysis. On the other hand, however, the more that emerging technologies such as wireless networks, broadband media, and home satellite systems become the norm for "global communications," the more naturalized and commonsensical the systems view of the world seems.[7] We can put this another way: systems-based views always seem to be ubiquitous, but never fully accepted, never quite fully naturalized. Taking a look at "historical" writings on systems theory, one of the first things to note is the fashionableness of the term and its general concepts. Thus, Ludwig von Bertalanffy, writing about a "general system theory" in the 1950s and 1960s, expresses some anxiety about the fact that suddenly there seem to be "systems everywhere" in disciplines as different as biology, psychology, communications, and sociology.[8] Such comments are not unlike those expressed more recently by cognitive neuroscientist Francesco Varela, who states, quite plainly, that "there is a reality of life and death, which affects us directly and is on a different level from the abstractions."[9]

Concerns about the "popularization" of concepts related to systems theories arise alongside an interest in making those concepts accessible to a wider, nonspecialist audience. If Bertalanffy puzzles over the conflation of his general systems theory with Norbert Wiener's science of cybernetics, so does Varela caution against the application of the notion of "autopoiesis" to extrabiological domains.[10] Much of this is owing to the interdisciplinary character of systems theories, which have be applied in fields such as economics, sociology, molecular biology, cognitive science, and business management. This broad reach of systems theories has found its most recent exemplar in a series of "pop-science" books published from the 1990s to the present, which concern an equally broad notion of "complexity." Books written by scientists or by science journalists covering "emergence," "complexity," "chaos," "self-organization," and other buzzwords have dominated mass-market science publishing, perhaps even eclipsing books on biotechnology or the human genome.[11]

More than one author has made the connection between systems views of the world and technologies such as the Internet, but often the connection is made in an over-optimistic, hurried way that belies the problematics in conflating the scientific, the cultural, and the political.[12] For instance, in an otherwise fascinating, far-reaching non-specialist book on self-organization and networks, Stuart Kauffman ends with a chapter aligning systems-based views of biological phenomena with an inherent, "democratic" order in social systems.[13] Other authors mime this same move, seeing in more holistic, global, inclusive scientific approaches to studying the world an innate benign quality, as well as a fundamental paradigm shift from traditional, analytic, reductionist science.

In order to investigate these problematics, it will be helpful to consider systems theory in historical perspective, and to enumerate some of its central tenets as outlined by Bertalanffy, and extended and modified by others. Bertalanffy's research implies a certain universality in adopting the specific phrase "general systems theory." Not unlike current research in complexity, Bertalanffy's systems theory seeks to identify the common patterns of organization across disparate systems. In so doing, it aims to establish the foundations for a science that is at once universally valid and constituted by particular conditions. As he states:

> there exist models, principles, and laws that apply to generalized systems or their sub-classes, irrespective of their particular kind, the nature of their component elements, and the relations or "forces" between them. It seems legitimate to ask for a theory, not of systems of a more or less special kind, but of universal principles applying to systems in general.[14]

It helps to understand the particular theoretical orientation of Bertalanffy's systems theory by briefly addressing two concurrent developments: information theory and cybernetics. A great deal of historical work has been done on both fields, and a significant amount of work has been done on their relationship to mid-century molecular biology, and the notion of a genetic "code."[15] What is important to note is that systems theory arises as much from a departure from information theory and cybernetics as its does from traditional science. Recall that, in Claude Shannon and Warren Weaver's formulation, information theory is the technical, quantitative analysis of the transfer of a message from point A to point B, a formulation that would establish some of the foundations for modern communications engineering. Working at Bell Labs during the 1940s, Shannon developed a model for technical communications that is largely followed today: sender → input → channel → output → receiver.[16] "Information" was here defined purely quantitatively, irrespective of content; what was of importance was the accurate transmission of a message along a communications channel. This notion of information—defined by quantitative measurement of transmission—may be said to undergird the Internet and computer networks. Although Shannon and Weaver do not make extensive use of the term *system*, it is implied in the technical potentials of their communications models.

In a similar vein, Norbert Wiener's research in cybernetics—a term he coined from the Greek root meaning "steersman"—placed equal emphasis on the role of "information," but differed from Shannon and Weaver in the way in which information was defined. For Shannon and Weaver, anything coming through the channel was information; the criterion for "good" from "bad" information was that the measurement of the message at point A was the same at destination point B. If it was not, the difference of the distortions was deemed "noise." For Wiener, not everything in a system was information; information was the result of a "choice" a human or a computer made from an array of options, or noise.[17]

Furthermore, whereas Shannon was primarily concerned with a model of unilinear communications channels studied in relative isolation, Wiener's more general vision involved multiple relationships of parts that constituted a whole. That whole could be living or nonliving, animal or machine—it mattered little for Wiener. Indeed, his book *Cybernetics* was subtitled *Control and Communication in the Animal and the Machine*. Wiener's cybernetic system was a series of loops in which output was fed back as input, so that a system could regulate its own performance. In classical cybernetics, the models of a ship (steering and rudder) and a thermostat are the most common examples. In each, an input (steering wheel; temperature dial) triggers an output (steer left or right; temperature up or down), the results of which are fed back as inputs (resistance of water; temperature already up or down). This process of feeding an output back as input Wiener called "feedback regulation" or just "feedback."[18] When feedback resulted in the system's return to a level state, the feedback was "negative"; when the feedback resulted in the system's further intensification of the prior state, the feedback was "positive."

In systems such as the thermostat, the process is automatic, not requiring human intervention, and in this regard it is noteworthy that a great deal of Wiener's work for the U.S. military during World War II was in automating ballistic trajectory tables for antiaircraft artillery—communication and control. Other research involved neurophysiological investigations into positive feedback conditions (excessive tremors, nervous system disorders). The basic model was the same, though the object of study was quite different. Finally, it was assumed in Wiener's model that the cybernetic system—again, living or nonliving—had an innate tendency toward a state of minimum energy consumption, an equilibrium, or a homeostasis. This was not simply a guess, but Wiener following the second law of thermodynamics, suggesting that closed systems naturally tend toward greater disorder or "entropy."[19] It was this tendency that, in large part, defined the kind of model developed in cybernetics.

In both the model of information theory and classical cybernetics, we can say that there is a configuration of a "system," in which component parts (technical in Shannon's case, more general in Wiener's case) relate together in such a way that a unit "information" is articulated, where "information" is taken as being equivalent to the "order" of the system. In Shannon's case, the order is established in the quantitatively accurate reproduction of a signal (or transmission of a message). In Wiener's case, order is established when the cybernetic system (living or nonliving) exists through continuous negative feedback loops, in which a state of minimal energy consumption or equilibrium is reached.

It was the postwar "buzz" surrounding cybernetics and the information sciences that prompted biochemists Francis Crick and James Watson to appropriate the metaphors of "information" and "coding" to describe their elucidation of the structure of DNA in the early 1950s.[20] In a series of articles, Crick would push the analogy further, so that, by the 1960s, the central concern of molecular biology would become "the coding problem"—how the four simple nucleotides of DNA were able to code for such a wide range of proteins.[21] It mattered little that Crick and Watson's appropriation of the discourse of cybernetics and information theory was inaccurate—their use was explicitly content-based and qualitative, while Shannon and Wiener's usage was primarily quantitative.[22] What was of primary interest to Crick, Watson, and others in the then-burgeoning field of "molecular biology" was the modes of understanding life at the molecular level opened up by the view from cybernetics and information theory.

Bertalanffy and Systems Theory

Concurrent with the work in information theory and cybernetics, Ludwig von Bertalanffy—a biologist by training—began to formulate a "general systems theory" that would take as its model neither communications technologies nor automated systems, but certain biological processes, such as cellular metabolism, membrane transport, and morphogenesis. Although various interpretations of "systems theory" would be applied to just about every domain except biology, it is important to note that Bertalanffy chooses biological models, both on the molecular level and at the "organismic" level. In fact, as he states, with regard to the "cracking of the code" in molecular genetics:

But in spite of—or just because of—the deepened insight attained by "molecular" biology, the necessity of "organismic" biology has become apparent . . . The concern of biology is not only at the physio-chemical or molecular level but at the higher levels of living organization as well.[23]

As we shall see, it is this tensioned space between these two levels—the biomolecular, bioinformatic body and the anthropomorphic, organismic body—that contemporary approaches in systems biology inadvertently raise as a central incommensurability. In Bertalanffy's formulation, systems theory is, first and foremost, an understanding of biological life that attempts to scientifically eschew the reductionism of molecular genetics, as well as the "closed system," quantitative approaches of information theory and cybernetics.

Keeping this in mind, there are three aspects to understanding Bertalanffy's systems theory, and how it was consciously a departure from the then-dominant scientific theories and practices. The first aspect is methodological, and it involves a move away from reductionist paradigms—in which an object of study is broken down into its component parts, and then the parts studied in isolation—and a move toward a more "holistic" viewpoint. Speaking about the aims of a general systems theory, Bertalanffy notes:

> While in the past, science tried to explain observable phenomena by reducing them to an interplay of elementary units investigatable independently of each other, concepts appear in contemporary science that are concerned with what is somewhat vaguely termed "wholeness" . . . in short, "systems" of various orders not understandable by investigation of their respective parts in isolation.[24]

However, well aware of the then-popular theories of vitalism—the notion that what defined biological "life" was something beyond the domain of empirical, scientific inquiry—Bertalanffy sets up one of the quasi paradoxes of systems-based theories to this day:

> The meaning of the somewhat mystical expression, "the whole is more than the sum of its parts" is simply that constitutive characteristics are not explainable from the characteristics of isolated parts. The characteristics of the complex, therefore, compared to those of the elements, appear as "new" or "emergent."[25]

While denouncing vitalist theories as "mystical," Bertalanffy also critiques the reductionist assumptions of traditional science. The "system" Bertalanffy attempts to describe is at once physical and metaphysical, its logic or organization inherent in the system itself, but also superseding the mere enumeration of systemic parts. From a methodological perspective, this holistic view of a system involves first identifying the system as such, and then articulating the relationships that will be foregrounded in the study of the system as a whole. From a scientific perspective, the main challenge of such an approach is to decide whether to exhaustively quantify the parts, so that their relationships are evident, or whether it is better to develop more qualitative, "global" analyses of systemic behavior. As we shall see, contemporary efforts in systems biology are split between these two decisions. For Bertalanffy, however, the holistic approach is explicitly posed against the methods of traditional scientific inquiry. "Holism" comes to stand in opposition to "reductionism," just as Bertalanffy's emphasis on the whole "organism" comes to oppose molecular genetics' emphasis on the component parts of DNA and proteins.[26]

Bertalanffy's holism is also positioned against the presuppositions in the physical sciences of his time concerning causality and teleology. Whereas classical physics and Newtonian mechanics pictured the world as a clockwork mechanism, fully reversible, fully determined, the infamous "three-body problem" and second law of thermodynamics first established the fissures in the reductionist model of linear causality and goal-directedness.[27] For Bertalanffy, the logical result of an analytic and reductionist approach is that elements can only be taken in isolation, and, as such, can only be conceived as interrelated in a singular, linear fashion.

Likewise, Bertalanffy repeatedly distinguishes his systems theory from information theory and cybernetics. It is interesting to note that, at the same time that Crick and Watson elucidate DNA's structure—and use the language of informatics to do so—Bertalanffy expressly counters this move by opposing the holistic "complex" of the organism to the reductionist "atomism" of molecular genetics. Although Bertalanffy notes that a great many of the processes in the biological body can be interpreted through the lens of cybernetics (feedback, input-output servo-mechanisms, homeostasis), there are also a great many other processes that do not easily avail themselves to the informatic approach (embryological development, the organism's apparent contradiction of the second law of thermodynamics).[28] In short, Bertalanffy suggests that, although the cybernetic view of the organism (e.g., genetic "code") is possible, it washes over a great deal of the systemic interactions of the parts as a whole.

These are, thus far, theoretical orientations, but how does Bertalanffy propose to carry out such analyses? The last methodological orientation is more specific, as demonstrated by the numerous differential mathematical equations throughout his texts. We need not go into the exact definition of differential equations to note that, from the beginning, Bertalanffy sets out to develop a flexible mathematics based on differences. The point here is not to debate the use of mathematical models itself, or the specific choice of differential calculus; rather, it is simply to note that, in the way in which differential equations imply a series measured over time, there is an emphasis on both static relations between components and systemic dynamics over time.[29] The "difference" Bertalanffy seeks here is not the difference of parts in isolation (one DNA sequence from another), but the difference of a system with itself, and within itself.

This methodological direction—and this is the second aspect of systems theory—sets the stage for Bertalanffy to articulate the implied universal characteristics of systems. What is a "system" for Bertalanffy? As he simply states, "A system can be defined as a set of elements standing in interrelations."[30] This is to say that the emphasis in systems theory is on "relations," and not on "things." In any grouping of elements, Bertalanffy states, a series can be ordered based on quantity (how many are there?), on type (what kind are they?), or on relations (which participates in this process?). The first two types of grouping are "summative," in that they isolate and abstract the elements according to parameters external to the specific context. The latter grouping— relation—is "constitutive," since it depends on the specific context and the relations between components that establish that context.[31]

In this sense Bertalanffy makes one of his most significant distinctions, that between "open" and "closed" systems. The primary difference between them is in the way in which systemic interior and exterior are configured. As Bertalanffy puts it, "An open system is defined as a system in exchange of matter and energy with its environment, presenting import and export, building-up and breaking-down of its material components."[32] Whereas closed systems are separated from their environment, open systems exist in constant interaction with their environment, exchanging matter and energy while maintaining their organization as a system. As Bertalanffy notes repeatedly, this distinction between open and closed systems serves for him to make ontological distinctions between living and nonliving systems, without resorting to what he terms the "mystical" characterizations of vitalism.[33]

Such open systems therefore exist in a qualitatively different type of dynamic continuum than closed systems, whose primary perturbations will be externally applied, and whose primary reactions will be determined by the closure of the system itself. This system-in-time is also a differential system, or a system whose components and relations are constituted by a variety of differences. Bertalanffy, using differential mathematics, identifies several types of dynamics in open systems: stable dynamics (where the differential between inflow and outflow is equilibrated), periodic dynamics (where patterns of regularity occur in cycles), and loop dynamics (where outflow feeds back into inflow, and vice versa).[34]

In addition to dynamics, there are several other general characteristics that, for Bertalanffy, describe open systems: growth of the systems' components (referencing laws of natural growth), competition between systems and within systemic regions (where one region of a system is expressed as the function of another region of the system), wholeness of systemic behavior (abetted in part by "progressive segregation" or differentiation of parts), and "equifinality."[35]

Finally, a third aspect of Bertalanffy's systems theory: As a way of filling out his theory, Bertalanffy uses biological models—not necessarily molecular biological models—to specify and further characterize systems theory as a theory of "open systems." Foremost in this specification of systems theory is a holistic view of the organism as opposed to the parts of an organism:

> If you take any realm of biological phenomena, whether embryonic development, metabolism, growth, activity of the nervous system, biocoenoses, etc., you will always find that the behavior of an element is different within the system from what it is in

isolation. You cannot sum up the behavior of the whole from the isolated parts, and you have to take into account the relations between the various subordinated systems and the systems which are super-ordinated to them in order to understand the behavior of the parts.[36]

As we shall see, such an approach generally informs current systems biology perspectives. A primary question in this view of open systems has to do with how the open system is open. In other words, given a system that is open with respect to its environment, is there any characteristic that remains constant, such that the system can be identified as a system? To be sure, mere aggregations of elements do not automatically form a system in a constitutive sense. How, then, is a system able to be identified as a system? To this question Bertalanffy suggests a halfway point between closed systems and disordered aggregations. What is important in this formulation is the connection he makes between the concept of open systems and his concept of the "steady state" (*Fliessgleichgewicht*):

> Conventional physics deals only with closed systems, i.e., systems which are considered to be isolated from their environment . . . However, we find systems which by their very nature and definition are not closed systems. Every living organism is essentially an open system. It maintains itself in a continuous inflow and outflow, a building up and breaking down of components, never being, so long as it is alive, in a state of chemical and thermodynamic equilibrium but maintained in a so-called steady state which is distinct from the latter.[37]

If the examples of biology constitute the paradigmatic instances of open systems— that is, if "living systems" are the model for open systems—then open systems are conceived of as a particular organization or relation of components in interaction with their environment through the continuous flow of matter and energy. As Bertalanffy suggests, the biological examples of selective cellular regeneration, biochemical metabolism, active transport across membranes, and embryological development or morphogenesis all provide instances of open systems that, despite their being involved in constant change, are still able to maintain their organization as systems. It is this phenomenon, which Bertalanffy describes as the "fundamental mystery of living systems," that constitutes the steady state.[38] The steady state is not so much a particular "thing," component, or substance, but rather, what it is itself is conserved through the flux of change in open systems.

This relationship between open systems and steady states is, for Bertalanffy, what distinguishes living from nonliving systems (be they chemical or mechanical nonliving systems). One of the basic assumptions in Bertalanffy's organicist or holistic approach is that there does exist this ontological distinction between living and nonliving systems. Mechanistic models of living systems—from Descartes's clockwork body to Wiener's cybernetic systems—are based on the model of classical physics, that is, on the model of closed systems. Bertalanffy directly critiques the mechanistic vision of living systems because it does not address three issues: the generation or origination of the living system, the adaptive modes of self-regulation of a living system, and the dynamic exchange of matter and energy of living systems as open systems.[39]

Even further, living systems as open systems are, according to Bertalanffy, to be distinguished from Wiener's cybernetic model on several accounts: first, regulation in the system is, in cybernetics, based on preestablished, fixed relationships between components; second, causality in cybernetic feedback loops is both linear and unidirectional; third, and most important, "typical feedback or homeostatic phenomena are 'open' with respect to incoming information, but 'closed' with respect to matter and energy."[40] This latter distinction means that the only thing that enters the closed cybernetic system is the feedback, or the difference between input and output, formulated by Wiener as "information."

There is much to explore here if we consider the appropriation of information tropes in molecular genetics, and Bertalanffy's explicit critique of genetics' reductionism. It is clear that the early formulation of molecular genetics' "central dogma" was anything but an open system in the way described by Bertalanffy. However, more recent investigations into alternative approaches to gene-centric studies of the genome have suggested that there may be some middle ground between the informatic interpretation of the genome and Bertalanffy's anti-informatic organicism.[41]

In short, Bertalanffy seems to want to suggest that materiality is immaterial from the systems viewpoint: "If open systems . . . attain a steady state, this has a value equifinal or independent of initial conditions."[42] The question here is, does matter matter? To what degree are the criteria of living systems constituted by material substrates? On the one hand, Bertalanffy resists fully abstracting living systems by an emphasis on their identity as systems in constant flows of matter and energy (as opposed to closed systems). But, from his holistic or organicist viewpoint, this attention to the context of matter is recuperated into a metalevel, at which something called the "organization" takes precedence over particulate material instantiation (we shall see, something analogous happen in theories of autopoiesis). Matter seems to constitute, but not identify, the living system as a system. What remains is a pattern of organization, and what changes is the flow of matter and energy through the system. In Bertalanffy's formulation, the latter seems to enable the former, expending itself in the process. The issue that general systems theory brings up is whether a systems-based approach on relations implies a divestiture of the constitutive context of materiality.

Systems Biology and Somatic Variability

Recall that, for Bertalanffy, the "organismic" model in systems theory explicitly excludes the molecular genetics notion of a genetic "code." Aligning molecular genetics with cybernetics and information theory, Bertalanffy takes genetics as part of a long tradition of mechanistic explanations of the human body. As he states, "the most recent development is in terms of molecular machines . . . it is a micromachine which transforms or translates the genetic code of DNA of the chromosomes into specific proteins and eventually into a complex organism."[43] The problem with such interpretations for Bertalanffy is their inability to make qualitative distinctions—we might even say "affective distinctions"—in a given state of the organism:

> What is the difference between a normal, a sick, and a dead organism? . . . One DNA molecule, protein, enzyme or hormonal process is as good as another; each is determined by physical and chemical laws, none is better, healthier or more normal than the other.[44]

While wanting to point out the inability of purely quantitative, informatic approaches to the organism (a binary genetic "code," protein "machines," a genetic "program") to address qualitative states, Bertalanffy also wants to sidestep the anthropomorphizing issues that a certain vitalistic notion of the organism would raise. To this end, he points out several distinctions between the molecular genetic notion of "codes" and the systems theory model of the "organism." Primary among these is the distinction between open and closed systems. Although the cybernetic model in molecular genetics may transmit information, it says very little about substance and structure; it thus conceives of a system that is thermodynamically closed (as in cybernetics and information theory), while at the same time allowing for the passage of genetic information.

Indeed, the very notion of gene expression and genetic "regulation" in contemporary genomics implies this dual foundation of information transfer as a means of establishing homeostasis. In Bertalanffy's model, the organism is situated at the nexus between a closed information system (as cybernetics might put it) and an open material-physical system.[45] A thermostat receives a signal about the room temperature from its inputs, but it does not actually receive the room itself; similarly, in steering a boat, the rudder sends a signal of water resistance to the steering wheel, but it does not send the water itself. A certain type of abstraction occurs, according to Bertalanffy, in cybernetics, in which information may flow quite abundantly, but not matter or energy. Quite the opposite case for organisms, which for Bertalanffy are principally defined by their being open systems, in constant interaction with their environment on a material and energetic level—matter and force. The "wrong turn" that molecular genetics makes, from this systems theory perspective, is that it confuses the abstraction of information-as-signal with the exchange of matter and energy in living organisms. The result is a somewhat paradoxical notion of DNA as a "code." However, without being a molecular biologist, one can immediately intuit the notion that DNA is not just a code, but a molecule, a complex of matter and force. The common processes of transcription (from DNA to RNA) and translation (from RNA to amino acid chain) are not just signals, but interactions of matter and force—this has recently become painfully evident in proteomics, where the study of protein folding has brought the question of structure back onto center stage.

From the systems theory perspective then, the molecular genetics view of the organism is both reductive (in that it breaks down the organism into its component parts) and exclusively analytic (in that it studies isolated components as closed systems, though they may transfer "information" to maintain homeostasis). Running counter to Bertalanffy's notions of a synthetic view of open systems whose elements interact in a steady state, the molecular genetics view of codes, programs, and instructions is therefore incommensurate with systems theory.

Although this may have been true for postwar molecular biology and cybernetics, current developments in biotechnology and genomics research are suggesting that the gap identified by Bertalanffy—between systems theory and molecular genetics—may be filled by the introduction of new computing technologies. Broadly termed "systems biology," and encompassing a number of university-based and independent research labs worldwide, systems biology addresses the gene-centric reductionism of traditional molecular genetics through highly specific methods of technology development. However, whether such approaches can offer an authentic alternative to genetic reductionism in its scientific and social aspects is something still left unaddressed by systems biology research.

How does systems biology relate to "classical" systems theory? There are several points of congruence in their general outlook. We can take the Institute for Systems Biology's (ISB) proof-of-concept study for systems biology as an example.[46] The ISB's article on sugar metabolism in yeast cells—GAL, or "galactose utilization"—makes use of a model organism and a well-known phenomenon to showcase the "difference" that systems biology presents to conventional biotech research. In this study, ISB researchers performed a systems-based four-step experimental protocol in their study of GAL. The initial step involved the development of an initial systems model based on preexisting knowledge of the GAL system. This included the definition of all components and processes relevant to the particular pathway of GAL metabolism in the yeast cell. The next step included the systematic perturbation of specific pathway relations through genetic and environmental modifications. Genetic modifications included partial deletion or overexpression of genes, and environmental modifications included growth or decrease in galactose levels in the medium. A third step was then to make use of computer technology (specifically, microarray and genomic techniques for detecting RNA and protein quantities) to integrate the experimental data with the initial systemic model. Finally, this integration of initial and experimental data would prompt the formulation of hypotheses to address phenomena not explained by the initial model. Additional protocols for perturbation could then follow upon this, repeating the same general steps (Figure 19).

It should be noted that the ISB chose galactose utilization in the yeast *Saccharomyces cerevisiae* because it represents a paradigmatic example of a "genetic regulatory switch," in which the processing of galactose only occurs, or is triggered by, the presence of galactose, as well as the absence of other molecules ("repressing sugars" such as glucose). Broadly speaking, cellular metabolism involves the coordination of multiple elements in "digesting" molecules and converting them from a nonuseful form to a useful one (energy, structural elements, cellular communications). The GAL process is an instance

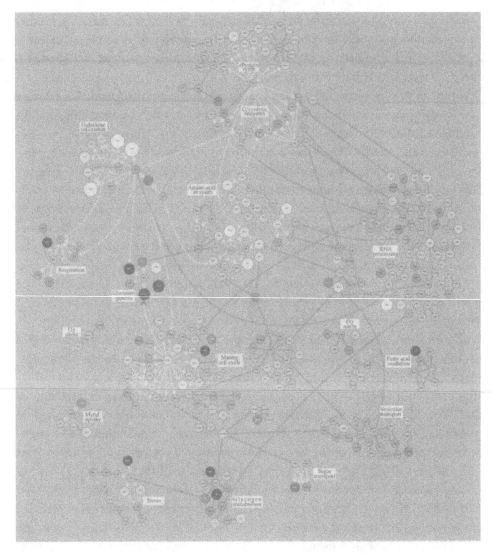

Figure 19. Network diagram showing metabolism of sugar in the yeast cell, from research by the Institute for Systems Biology. Reproduced courtesy of *Science*.

of a particular type of cell processing a particular type of molecule in order to extract energy from it.

Briefly, the ISB's findings both reinforced already-existing knowledge about *GAL* metabolism in *Saccharomyces cerevisiae* and produced unexplained findings that, it is implied, extend from the ISB's systems-based approach. For example, under normal conditions, the presence of the gene *gal7* is known to play a role in regulating the level of *GAL* enzymes in relation to the presence of galactose (when galactose is present, the enzymes to digest it are produced). The ISB found that when the *gal7* gene is deleted entirely, repression of *GAL* enzymes in the presence of galactose still occurs. The ISB researchers note that in the absence of *gal7*, GAL metabolites build up in excess, eventually reaching a toxic level for the cell. One systems-based hypothesis that they derive is that the buildup of *GAL* enzymes in the absence of *gal7* regulation triggers a lateral response by other *GAL*-regulating genes, which then immediately lowers the levels of *GAL* enzymes in the cell. This lateral influence—a kind of backup plan—is something that depends on a view of cellular metabolism that extends beyond an isolated, single gene-protein causal chain.

To return to our question: how do systems biology experiments such as the ISB's relate to Bertalanffy's systems theory? There are three basic correspondences between systems biology and Bertalanffy's systems theory. One is an emphasis on "relations" rather than "things." In biotech research this means an emphasis on biological "pathways" and metabolic "networks" rather than individual "genes." Rather than searching for the gene or genes in a yeast cell that are responsible for the metabolism of the sugar galactose, the systems biology approach will consider metabolism as a complex of heterogeneous elements (DNA, transcription processes, enzymes, signaling pathways) that function together in a specific way in converting galactose into cellular energy. These elements and their relations do not cause or produce the energy; rather, they are the galactose metabolism (recall Bertalanffy's distinction between summative and constitutive relations). The elements are not isolated, static "things," but are indissociable from the actions that constitute them. In this sense, the "elements" of a system may include diverse phenomena, "things" such as DNA, and processes such as posttranscriptional editing.

Another commonality between systems biology and classical systems theory is an emphasis on "perturbations" rather than "mechanisms." In contrast to the mechanistic approach of cybernetics and information theory, Bertalanffy proposes an analysis of the networks of relations between elements as generative of larger, "organismic" systems behavior. Likewise, systems biology generally focuses on first articulating a system for study, and then proceeding by methodically introducing perturbations to selected pathways—in the same example of yeast cellular metabolism of galactose, an understanding of the processes from systematic experimental modifications of the network. Perturbations—also known as "knockout" experiments—may either have no effect on the specific phenomenon being studied (an example of Bertalanffy's equifinality) or have a moderate effect, processing the sugar, but incompletely, inefficiently, or with side effects; or it may perturb the metabolism at a constitutive level, making the specific relation being perturbed an essential relation. Each perturbation is an experimental step, revealing characteristics about the relationships between elements rather than about isolated components. As Bertalanffy states, multiple pathways, "equifinality," and nested systems all testify to examples from the biological domain that exceed linear cause-effect chains, a result of the organism existing as an open system. Systems biology research, while still articulating the "system" itself according to traditional molecular biology (that is, the "boundary" to a system must first be articulated in an experiment), shares this emphasis on multicausality—though, as we shall see, it remains somewhat vague on its relation to the question of medical therapy.

Systems biology and systems theory also share a broad epistemological outlook, which is based on "patterns" rather than "identities." The view of relations-not-things, and perturbations-as-change, is undergirded by observing practices that highlight the systems-wide behavioral patterns instead of the differentiating process of identifications. For systems biology, this translates into both a coordinated management of biological data across different substrates (DNA, gene expression, proteins, enzymes, signaling) and technologically enhanced modes of representing that data in congruence with a systems view (computer-generated and calculated diagrams and topologies). The aforementioned study of yeast cellular metabolism of galactose involved techniques and technologies from an array of subfields in biotechnology: biochip diagnostics and gene expression profiling, high-throughput genomics and proteomics, and bioinformatics. As can be seen in many related examples in biotech research—principally bioinformatics—this "pattern epistemology" is in many ways indissociable from a high-technology interpretation of the biological domain.

Such an approach is more than a purely theoretical orientation. For a number of years, the search/discovery of isolated genes related to a given disease has been a mainstay of medical genetics approaches to treating disease (e.g., vector-based gene therapy). Likewise, the techniques for discovering and manipulating such genes have been a part of the biotech industry and patenting policies in the United States. The claim that most biotech companies make concerning patenting is that such practices are a form of investment for the company, in order that it might seek returns on the expenditures needed in the research and discovery process. Gene-centric approaches also tie patenting practices to the pharmaceutical industry, where "gene targeting" is a standard technique for helping to bring potential drugs or therapies to clinical trials. The notion of gene targeting in the drug discovery process has wider implications for molecular biology research. Although most researchers (and CEOs) will likely agree that any genetic condition is "caused" by a complex of multiple genes and proteins, the very fact that the human genome logically precedes the human "proteome," and that analyses of gene sequences dominate bioinformatics tools and research, is indicative of the still-pervasive presence of gene-centric approaches to understanding "life" at the molecular level. In short, the traditional molecular biology emphasis on "things" is as much part of the scientific tradition as it is part of the still-new tradition of the biotech "industry."

However, there are also important differences between Bertalanffy's systems theory and the systems biology perspectives mentioned here. A primary one is the relation to technology and technical tropes. Bertalanffy explicitly critiques the mechanistic outlook of molecular genetics and cybernetics, whereas systems biology appears to see no incongruence between the use of high technology and the affirmation of "global," holistic approaches to understanding biological phenomena. Bertalanffy's critique was based on the assumption that both cybernetics and molecular genetics conceived of closed systems, a viewpoint that, he points out, is commonsensically opposed to the way in which organisms actually exist. Thus, for Bertalanffy, not only do the tropes of mechanism frame the organism as a closed system (existing in reversible time, operating through linear causality, tending toward homeostasis), but the phenomena he points to (such as metabolism) are ontologically distinct from the thermostats of cybernetics.

For systems biology, taking a systems approach to the biological domain does not imply a renunciation or a critique of molecular genetics' informatic interpretation of the body (as it did for Bertalanffy). This is in part because of the widespread, even "natural" assumptions that DNA is a "code," that the genome is a "program," and that, on a philosophical level, the Human Genome Project's uncovering would be akin to a decryption of a coded message. However, beyond the naturalized status of the informatic view of biology, systems biology also replicates molecular biology's confusion of information-as-signal (as in cybernetics) and the exchange of matter and energy, which, for Bertalanffy, defines living systems. The twist, however, is that, with the aid of new computing technologies, systems biology rematerializes the exchange of matter and energy in novel ways. This is what we have been referring to as "biomedia": the technical recontextualization of biological components and processes. Unlike genetic engineering, systems biology does not manipulate, engineer, or modify the organism itself; but it does use informatic technologies to situate and thereby articulate biological phenomena in a particular way. The use of computer technologies geared toward database management, multitasking, and computationally intensive analyses of data strings are all technical contexts within which the "natural" phenomenon of yeast sugar metabolism takes place.

But this recontextualizing does not happen in just one way, and this is what differentiates systems biology from other related fields, such as bioinformatics and genomics. In the ISB's research, for instance, a number of different technologies are, in effect, "plugged into" each other. It is important to note that, although a researcher may utilize DNA microarrays, gene sequencing machines, and bioinformatics tools, rarely are such technologies used singularly, in one experimental gesture. Most often, the technology, being highly specialized, is used to generate data, which may then be combined together to give an overall picture of the phenomenon under study. Thus, although the technologies themselves are rarely plugged into each other, the data they generate are. The human genome mapping effort, for instance, makes use of automated gene sequencing computers, which are very different from the on-line software applications used for protein secondary structure prediction, which are in turn different from the range of biochips used in gene expression profiling. What the ISB's research has done is to integrate these technologies into a single systems analysis package. Because each technology has specializations, such an integration works well with the polydimensional challenges of a systems approach.

However, we may again ask about the difference in these heterogeneous technologies plugged into each other. Specifically, we can note that many of the computer technologies currently used are split along a "sequence-structure" divide. This essentially breaks down to whether a technology is used for studying biomolecular sequence (DNA, RNA, protein code, SNPs, ESTs), or whether a technology is used for studying biomolecular structure (signal transduction, protein folding, chromosomal structure). Even within a field such as bioinformatics, this divides software tools into those that analyze sequence (gene finders, sequence alignment applications) and those that analyze structure (protein folding coordinates, 3-D molecular modeling). The sequence–structure relationship is not new, but can be said to accompany the emergence of molecular genetics itself. When James Watson and Francis Crick elucidate DNA's structure, they do so through the lens of informatics. When François Jacob and Jacques Monod study the role of enzymes in regulating gene expression, they do so by articulating a genetic "program" that operates a genetic regulatory "switch."[47] In a sense, the way in which the sequence–structure relationship is configured can be said to define molecular genetics: sequence literally "in-forms" structure, in that DNA in-forms proteins in the same way that genotype in-forms phenotype.

In systems biology, the intersection of informatics and "organicism," or closed and open systems, brings to the fore issues that have been a part of molecular biology for some time. In bioinformatics, a common research protocol is to begin with sequence analysis (searching databases for homologies, performing sequence alignment), which provides data needed for more refined structure analysis (using sequence for structure homology, performing structure prediction based on sequence). Systems biology, by contrast, approaches biological phenomenon in a more nonlinear, polydimensional manner, simultaneously running sequence and structure analyses that are plugged into each other, and whose output forms an aggregate of different biological data types. Recall that the ISB's experiment on yeast cellular metabolism outlined four experimental protocols: identification of components and relations in a pathway, perturbation of pathway components, integration of systemic responses with the initial model, and hypothesis revision. At each step of this process, a systemic model is reconfigured, modulated, extended, or revised altogether; that is, the ISB's approach begins and ends with systems, or a systems-based view of the biological phenomenon under consideration. This is in contrast to more gene-centric research fields such as genomics, functional genomics, proteomics, or rational drug design, in which one begins from a model describing particular, relatively isolated, gene-based components, from which extend various products such as RNA or proteins.

Systems biology research distinguishes itself conceptually by two means. First, it adopts the computer programming approach of "versioning" its systems. The ISB's experimental protocol consisted not of a single yeast cell that was then allowed to run its course undisturbed. Rather, several cellular metabolic systems were devised, each one defining a particular perturbation to a particular pathway component. For instance, one set of yeast cells were perturbed by the partial deletion of gene segments that are known to correspond to *GAL* genes that express the synthesis of a galactose-metabolizing enzyme. Gene segments corresponding to transport, protein synthesis, or regulatory functions were individually perturbed in separate "versions" of the yeast cellular metabolic system. Likewise, environmental perturbations—such as the increase or decrease of galactose—were also included in separate versions.

This versioning of the biomolecular body leads to a second distinguishing characteristic, which is that each version runs "in parallel." In computer science terms, "parallel processing" involves the utilization of more than one processor chip in a computer system, with each chip performing specific computational tasks independent of the other processors, but working together as a single computer. In a like manner, the ISB's experimental protocol takes each version of the yeast cellular metabolic system as a processor; each system "runs" its own perturbation, and each is also an identical system, a part of a larger unit, whose output data is integrated in research.

These two characteristics—versioning of the biomolecular body, parallel processing biomolecular systems—culminate in a kind of "parallel corporeality," in which the yeast cellular metabolic system simultaneously exists in several different ways, each mode of existing defined by a specific systemic perturbation. These are also, however, corpo-realities, in the sense that the ISB's experiment is different from a simulation. Qualitatively different cellular systems are created by their being enframed, contextualized in a certain way as biomedia (excision of part of a gene; modification of environment). The living systems generated are different from those of genetic engineering, pharmacogenomics, or transgenics; they place less emphasis on the direct manipulation of the system's components and more emphasis on the technologies of recontextualizing the living system so that it can continue to function (naturally, biologically).

Perturbed Bodies

In aspects of systems biology such as versioning and parallel processing, a notion of how a "living system" is constituted through the systems theory framework is made apparent. As we have seen thus far, systems biology arises, in part, as a result of the deluge of data generated by the various genome mapping efforts. Once all the data has been gathered and archived, the question of what all the data means is raised. Systems biology—in the examples of the ISB and other like groups—offers a predominantly technological response to this question. From such a perspective, the question of biological "signification" can only be answered through new computing technologies that are capable of handling the extensive versioning and parallel processing required to manage—and articulate—this "biological data." Gene expression profiling, high-throughput sequencing, and computational protein structure analysis are key aspects to this approach to molecular biology. Genomics, proteomics, bioinformatics, and computer science are all enlisted in the effort. Along with this is also a unique approach that focuses not on "things" but "relations," a key aspect of a general systems theory as laid out by Bertalanffy. This emphasis on relations over things runs counter to both classical molecular biology (in gene-centric approaches) and the biotech industry (in the emphasis on gene targeting and drug discovery). In this sense, systems biology would appear to offer an alternative to gene-centric, reductionist approaches, through its particular design of a biotechnical context—this "biomedia" turned inside out, so that the context transforms meaning.

But maintaining this alternative approach is quite different from initially proposing it. The philosophical implications raised by the data generated by systems biology point to a notion of biological life quite different from how it has been traditionally understood. In particular, the implications of systems biology point to a central tension between "system" and "body," or, put another way, between "net and self."[48] The primary question that systems biology must address is whether a shift in methodology, techniques, and scientific orientation also entails a shift in philosophical outlook (including ethical and epistemological shifts). We can pose this problem another way: If the models of biological life in molecular genetics—genetic codes, genomic programs, genetic switches—are just as much cultural and philosophical in their reach as they are scientific and technical, then it would seem that any practice that is a conscious alternative to such models would also entail such "extra scientific" shifts.

There are a number of thorough studies of the practices, rhetorics, and "cultural icons" of DNA-as-code.[49] What might be the broader implications of a shift toward a systems biology? If we take the principles of systems biology to their logical conclusion—open systems, relations, patterns, pathways, disequilibrium, perturbations—we are led to a model of "the body" quite different from the anatomical, anthropomorphic body that has been a mainstay of Western medicine since Vesalius.[50] The body of systems biology is not anatomical in that it does not presuppose a whole that is constituted by its parts in a fixed, functional manner. However, we are also led to a model of the body that is equally different from the bioinformatic body of DNA and molecular genetics; for, while molecular genetics displaces the anatomical hierarchy of whole and parts, it generates the same hierarchy at a different level—phenotype–genotype, and the reductionism of gene-centric approaches to studying biological phenomena. In systems biology, the body is not "genetic" in the sense that it grants no priority to DNA or genes, perferring to see them as one component among many other components and processes. In molecular genetics, the "code" of DNA comes to play a wide range of roles, from master plan, to instruction manual, to dictionary, to source code or software. However, ideologically there is always the familiar human subject at its end, at the point where abstract, informatic genotype culminates in and in-forms the phenotype of the self. Because of—or in spite of—popular notions of "genes for X," this dependency on there being a human subject, a self, at the end of it all is, in many ways, a fundamental property of the linkages between genetics, medicine, and society: we may be constituted by biological information, but, in the end, that data is nothing without its being realized in the self.

It is interesting to note that systems biology says very little about the status of the self in relation to the myriad networks it materializes in its research. No doubt much of this is owing to the fact that, as an alternative approach, it is still relatively new and also often marginalized amid discussions of genetic drugs and gene therapies. Yet, what remains unarticulated thus far in systems biology is this connection between the technoscientific and the philosophical-ethical. Bertalanffy too addresses this tension between system and body, net and self. His response, however, is to attempt to recuperate modern notions of subjectivity and liberal-humanist notions of individuality into his systems viewpoint:

> Such knowledge [from general systems theory] can teach us not only what human behavior and society have in common with other organizations, but also what is their uniqueness. Here the main tenet will be: Man is not only a political animal; he is, before and above all, an individual. The real values of humanity are not those which it shares with biological entities . . . but those which stem from the individual mind.[51]

In philosophical-political terms, Bertalanffy inadvertently outlines one of the main tensions in systems theories: the negotiation between processes of identification and practices of distribution (system–body, net–self). For Bertalanffy, as in systems biology, the human subject is an organism, and the organism is defined by its systemic properties, and this defining is guided by principles of pattern rather than singularity, relation rather than identity. A first step in any systems-based approach is to articulate the system, that is, to demarcate a boundary. The second step is to complexify the first step by articulating the components and relationships that constitute that boundary. The tension that systems biology brings to the fore is whether this second step may in some sense fold back on the first, making the first step of identification biologically irrelevant.

Of course, this does not happen in systems biology research, and it should be reiterated that we are taking a certain tension to its extreme end point. What is "the body" in systems biology? We can begin by suggesting that the body is first and foremost a "system," implying all of the principles that Bertalanffy describes (its being open, maintaining a steady state, having elements as relations, exchanging matter and energy). Second, this system is not merely an abstraction, but is materialized in biological— and biotechnical—contexts. We could simply say that systems biology was producing representations or simulations, were it not for the fact that the patterns and relationships are contextualized in ways that make pattern and relation indissociable from biological materiality. Third, the body in systems biology—as both pattern and matter, relation and substance—is irreducible to any general model that takes the system as a simple building block for a larger unity. This is a contested point in systems biology, for although Bertalanffy emphasizes a holistic, "organismic" viewpoint, it remains unclear throughout his texts exactly how the distributed model of open systems is to be reconciled with the holistic model of "the individual."

These three characteristics define what we might call the "biomolecular body" in systems biology. However, "biomolecular body" is understood here in a sense quite different from its conventional notions in molecular biology. It is not the protein-specific body, it is not the "central dogma" molecule of DNA, and it is not even the decrypted code of genetics.[52] The notion of the "operon" model in genetic regulation is closer to systems biology, emphasizing as it does processes over structures and substances, and incorporating recursion (enzymes produced by DNA act on DNA to produce enzymes), distribution (polygenic factors), and multiple layering (signal transduction cascades).[53] The biomolecular body of systems biology is distributed (emphasis on nodes and relationships), in parallel (multiple intersecting processes), dynamic (based on perturbations to pathways), and focused on relations over substance/structure (pathways, not matter or form).

The biomolecular body is not anatomical or anthropomorphic, but neither is it cellular or genetic. Taken to its logical extreme, it is "system" opposed to "body," if by the latter term we mean a singular unity separated from its environment—the locus of the "self." The processes that systems biology groups such as the ISB study do not form building blocks for a unified whole (the organism, the self); rather, they form level-specified systemic functions that operate irrespective of the necessity to form larger, molar formations. It is in this sense that we can suggest that the biomolecular body is also an "inhuman" body. It is a set of patterns and relations that, though it may be embodied in our body, in the body of the self, does not constitute the self. To take a simplistic example, the notion of the "central dogma" in postwar molecular genetics is based on the constructionist, parts-make-up-the-whole approach, which is not unlike anatomical science: DNA makes RNA makes proteins, and proteins make us, was the central dogma's claim. The "purpose" of DNA is therefore to constitute the self, molecule by molecule, organ by organ, part by part. The biomolecular body in systems biology takes a different route. It rarely mentions the self at all, not out of any disregard or ethical oversight, but rather because, from the "perspective" of the biomolecular body, the self is irrelevant. What is relevant are the relations between components and processes, and their potential perturbations. Somewhat bizarrely, we could even say that systems biology asks the question: what is the phenomenology of the biomolecular body?

However, although systems biology opens up this new domain, it ends up foreclosing it in the question of concrete solutions to medical problems. Although it offers a systems-based technology design environment, it ends up referencing drug development and gene therapies as its primary contribution to medicine and biotech, the same as offered by the more mainstream efforts in genomics and proteomics. Currently, medical applications of biotech research are dominated by gene-centric models: drug discovery ("rational drug design," "pharmacogenomics"), diagnostics (DNA chips), and genetic therapies. Thus, medical genetics problems are framed in a particular way: that genotype is expressed in phenotype, and phenotype expresses itself in the patient/subject; that is, the medical genetics view is that the whole is the sum of its parts (what Bertalanffy calls an "additive" perspective). The person is affected by a disequilibrium that is genetic in nature; therefore, the solution is to identify the source of the affection (using DNA chips for genetic screening), and to directly counteract it, either through the design of custom-tailored drugs or through the introduction of counteracting molecules (antibodies, antisense therapies, vector-based gene therapies). The tension in the biomolecular body—system and body, net and self—is here reproduced in the particular context of biomedical application. It may then be said that the medical genetics paradigm holds a particular view of the subject, the self, the patient that is incommensurate with the biomolecular body in systems biology.

Systems biology operates through a lens that is "inhuman." It aims to apply its modes of knowing to problems that are specifically designated as "human" (as in the case of medical genetics). We can be more specific by asking what a body is in systems biology. A body for systems biology is a network, a set of relations between components through which perturbations may occur along one or more pathways in the network. What is the body in medicine? The body is a combination of anatomical views (anthropomorphic constructionism) and genetic views (genotype–phenotype, gene-centric), both of which maintain the hierarchy between levels and processes (the molecular only serves the molar; the end goal is the person, regarded biologically as the sum of his or her parts).[54]

Inhuman Bodies?

Systems biology has begun by addressing a technical question, but in the process has raised issues that are both technical-scientific and philosophical-ethical. The characterization of the biomolecular body as "inhuman" is not a moral description; *inhuman* does not necessarily imply *antihuman* or *nonethical*. It does, however, suggest that a radically "other" model of the body emerges from the predominantly technical orientations of systems biology, a model we have been calling the biomolecular body. The biomolecular body is articulated in systems biology practices in such a way that the philosophical-biological models associated with Western biomedicine and molecular genetics cannot accommodate it without significantly reducing the characteristics that qualify it as a living system. Because the biomolecular body in systems biology is neither anatomical-anthropomorphic, nor genetic-informatic, it also requires a theoretical understanding which would address the exact ways in which it departs from current paradigms.

As a way of elaborating the theoretical effects of the biomolecular body, we can consider two examples of interdisciplinary research that, each in its own way, attempts to understand the implications in a systems-based view of biological life. To be sure, many examples could be added to these, and in this sense it is more accurate to speak of systems biologies in the plural. Although many of Bertalanffy's concepts can be found in the contemporary sciences of complexity and self-organization, no two instances share the same set of specific configurations; indeed, if systems biologies have something to contribute to systems theory, it is whether Bertalanffy's dream of a universal, "general systems theory" is a viable aim for the sciences of complexity.

Boolean Genetic Networks

A first example is that of "Boolean genetic networks" or "N = K networks" elaborated by complexity researcher and biologist Stuart Kauffman. In a series of books and articles, Kauffman has for many years been a proponent of the sciences of complexity, using the biological domain as his model for studying processes of self-replication and emergent order in enzymes, peptides, and networks of genetic regulation.[55] One of Kauffman's basic claims is that there are unique laws of complex self-organization in living systems, enabling them to retain order in a dynamic context, what he calls the "edge of chaos." Such a claim runs counter to much of evolutionary biology, in which the two natural processes of random mutation and natural selection are thought to be the guiding principles in the ontogeny of living systems. Rather than accept this reliance on randomness, Kauffman shows—through theoretical and empirical research—how models from the sciences of complexity can provide a theory of self-organization that demonstrates an inherent order and logic in the development of living systems, what he refers to as "order for free."

In addressing these laws of self-organization, Kauffman looks to chemical catalysis as a model for spontaneous order. "Catalysts" are molecules that speed up chemical reactions through a lock-and-key fit between molecular structures. "Autocatalysis" is the set of catalytic reactions that make the molecules of which the catalysts are formed a self-sustaining network of dynamic chemical interactions. As Kauffman states, "What I call a collectively autocatalytic system is one in which the molecules speed up the very reactions by which they themselves are formed."[56] He suggests that

> life is a natural property of complex chemical systems, that when the number of different kinds of molecules in a chemical soup passes a certain threshold, a self-sustaining network of reactions—an autocatalytic metabolism—will suddenly appear. Life emerged, I suggest, not simple, but complex and whole.[57]

What the model of collective autocatalytic systems demonstrates is that every molecule in a living system is related to a network of other molecules. Although discrete interactions can be identified and isolated, action, control, and agency are all distributed among the network. Once this distributed connection is established, a system can catalytically maintain itself through reactions that form the components that perform the reactions, a vicious circle that Kauffman calls "catalytic closure":

> I hold that life, at its root, does not depend on the magic of Watson-Crick base pairing or any other specific template-replicating machinery. Life, at its root, lies in the property of catalytic closure among a collection of molecular species . . . DNA replicates only as part of a complex, collectively autocatalytic network of reactions and enzymes in cells.[58]

As a way of illustrating such relationships, Kauffman proposes the "N = K network" or "random graph" experiment. In this exercise, which can be performed using buttons and thread, "nodes" (N = buttons) and "edges" (E = threads) are combined in various ways at each step. If N is much higher than E, then it is quite possible to isolate unconnected nodes. If E becomes closer to or greater than N, then we have a more thorough network. The trick is to identify the phase transition in E/N, from a series of disconnected small networks to a single whole network. A graph of this relationship shows that at a certain point there is a significant jump in the size of the cluster. The analogy is that when a large enough number of connections exists, a collective auto-catalytic reaction will occur.

Two scientific concepts inform Kauffman's theoretical research—one scientific and one technological. The first is the "operon" model of gene expression articulated by François Jacob and Jacques Monod in the 1960s.[59] Jacob and Monod used the informatic trope of a genetic regulatory "switch" to describe the ways in which DNA produces protein A, which in turn acts on another protein B, which regulates whether or not DNA will produce the first protein A. This is combined by Kauffman with Boolian logic circuits in modern computers. Boolean algebra, described by mathematician George Boole in the nineteenth century, states that any observed phenomenon can be reduced to a series of yes/no statements, which, in computing terms, can be translated into binary digits (yes/no; 1/0) and sets of binary conjunctions (AND, OR).[60]

Combining these two concepts, Kauffman performs the same graph experiment for "reaction networks" of metabolic activity in the cell. Here the nodes may be particular molecules (gene sequence, amino acid chain), and the edges may be intracellular processes (transcription, enzymatic action, membrane transport). In this model, the processes such as gene expression may be taken as being "on" (1) or "off" (0). The logic of the network is thus composed of a series of on/off switches, constituting a Boolean network, or network of Boolean instructions (AND, OR, NOR, XOR). The given processes, such as the transcription of RNA from a given gene A, can be either expressed OR not expressed, or it can be expressed if two other, separate genes—gene B AND gene C—are also expressed (a precondition statement). The same can also be said for the NOR (A is expressed if neither B nor C are expressed) and the XOR (and/or) statements (A is expressed if either B or C is expressed).

When the ratio of nodes to edges is significantly high, the diversity of the molecules is increased, such that the products of the reactions may become catalysts. When the number of reactions equals the number of nodes (at least one reaction per node; nodes have at least one edge), then collective autocatalysis occurs. There are three general results in this process. First, a greater diversity of molecules (nodes) equals a greater number of reactions (edges). Second, a greater number of catalysts produces a tendency toward phase transitions. Third, when the E/N ratio equals 1, then autocatalytic closure occurs.

For Kauffman, biological "life" emerges at the point at which autocatalytic closure takes place, because such systems are self-reproducing, self-maintaining, and sufficiently diverse (beyond the minimum threshold of complexity). Using this approach, Kauffman begins to build a model of an autocatalytic system using Boolean networks to model the interactions of genes in a genome. In this system, the nodes are individual genes or DNA sequences, and the edges are the particular processes under consideration (gene expression). The first claim Kauffman makes is that living systems can be autocatalytic without relying on the template-replication (unilinear) models of the genome:

> In short, there is reason to believe that autocatalytic sets can evolve without a genome. There is no separate DNA-like structure carrying genetic information ... With autocatalytic sets, there is no separation between genotype and phenotype. The system serves as its own genome.[61]

In addition to specifying the spatial orientation, the temporal factor needs to be taken into account—how each lightbulb will respond to signals from other bulbs. For instance, if both inputs are off ("oo"), then the lightbulb may switch from on to off or vice versa ("AND"), or it may switch if only one or the other is off ("OR"). When "run," this network will generally appear to be a bank of flickering lights. If the network size is small, then it is more probable that it will repeat each of the possible states over again—such a repetition is a "state cycle." State cycles can be large or small, depending on the "attractors" present in the system.[62] A very large state cycle will be so large that it is essentially unpredictable, whereas a small state cycle will enable the system to establish an equilibrium. Therefore, attractors can "trap" the system into state cycles, as one source of order.

Two factors dictate how Boolean networks will behave (behavior that may be chaotic, ordered, or at the "edge of chaos"): first, the number of connections or inputs (sparse connections result in ordered systems, densely connected results in unstable systems), and second, the control parameters (AND and OR are examples from Boolean math; other control parameters can exist such as NOR). Modulating or "perturbing" either of these (the number of connections, K, or the control parameters, P) can be a way in which a system maintains itself at the edge of chaos. These are "N = K networks," where K is the number of connections per node, and N is the number of nodes. The higher K is, the more complex the network becomes (and the longer the state cycles). Biases in the control parameters can have a significant impact on the Boolean network's ability to display complexity. Kauffman proposes studying such networks from the perspective of control actions.[63]

Boolean networks can display high degrees of order (state cycles that are small loops) and high degrees of disorder (large state cycles with so many variables that their totality cannot be measured). Kauffman's goal is to find a way to highlight the point between chaos and order—or the "edge of chaos"—where dynamical systems such as living organisms exist. Kauffman's model aims to show that, in many ways, genes are the least significant part of a living system such as a cell; in Kauffman's view, the reaction pathways are the elements that qualify a system as "living" (recall Bertalanffy's emphasis on relations rather than things).

Note that, working independently, Kauffman arrives at a space that is very similar to the systems biology research of the ISB discussed earlier. Like the ISB's experimental protocol, Kauffman's N = K networks emphasize relations over things, configure such a network as a field of potential perturbations, and from such studies propose the formulation of a model of the system. For molecular genetics, one of the implications of N = K networks is that, when they exist at the edge of chaos, they enable more complex interactions than direct gene-protein relationships. For instance, under certain conditions, a gene at one location may act as a promoter or silencer for a gene at a distant location in the network, which will then express a third gene region, which will then trigger the production of enzymes used in the regulation of a certain molecule.

What Kauffman's N = K networks add, however, is something latent in the ISB's approach, and that is the integration of informatics with molecular biology. Recall that, for Bertalanffy, molecular genetics mistakenly interprets "life" as the transfer of genetic "information" within closed systems (the model of cybernetics). Molecular genetics could never conceive of open systems, much less living systems, because of its reliance on the cybernetic model of the nonexchange of matter and energy. The ISB appeared to counter this claim in a purely pragmatic manner, opting to integrate high technology into the modeling of living systems at the molecular level. Kauffman takes this a step further, suggesting that, in examples such as collectively autocatalytic sets, the role of DNA is nonexistent. Kauffman's N = K networks not only reaffirm the systems-based approach to biological life, but, more important, they do so through an explicitly informatic model: that of Boolean biomolecular "software" (the application of Boolean algebra to gene expression) and the circuitry of genomic "hardware" (his simulation using lightbulbs and electrical circuits).

Despite this, Kauffman, like Bertalanffy, also runs up against the tensions between system and body, net and self. When addressing the larger philosophical-ethical aspects of this development of self-organization and complexity in biological systems, Kauffman resorts to a negotiated form of teleology. In short, he seems to simply take a new approach to achieve the same results; although he provides a theoretical foundation for understanding living systems, he does not question the parameters for defining such systems as "living" (reproduction, replication). In his countering of evolutionary biology's acceptance of chance mutations, Kauffman suggests that the "universal," "natural" laws of complexity demonstrate that biological life is not without a "deeper," extrabiological significance:

> the emerging sciences of complexity begin to suggest that the order is not all accidental, that vast veins of spontaneous order lie at hand. Laws of complexity spontaneously generate much of the order of the natural world ... organisms are not just tinkered-together contraptions, but expressions of deeper natural laws ... Not we the accidental, but we the expected.[64]

The negotiation made here results in an inhuman teleology. While denying the conventional teleological explanation of biological life ("life" as goal-directed, bearing inherent, "higher" significance), Kauffman also denies the explanation of biological life from the view of evolutionary biology (blind mechanisms running on chance mutations). In between are the networks, pathways, and autocatalytic sets that articulate, in very "inhuman" terms, the concept of an "other" logic. In effect, Kauffman's models demonstrate this implosion of self-organization with teleology, resulting in an interpretation of biological life as inherently ordered, or rather, as containing within itself an inherent, implicit capacity to balance itself on the "edge of chaos" (recall Bertalanffy's concept of the "steady state" in this context). On the one hand, we have complex networks from relatively few nodes and edges, the complexity resulting from their combinations, resulting in living systems poised at the "edge of chaos." On the other hand, we have a set of impersonal, inhuman regularities that, according to Kauffman, express an inherent, even benevolent, desire for order. Complex self-organization thus appears to be at once human (these regularities configure a biological teleology) and inhuman (they do not foreground the biological "individual").

Autopoiesis and Nonsubjective Expression

If Kauffman's Boolean genetic networks result in a theoretical confusion of systemic function and teleology, we may look to another, somewhat like-minded approach that also takes the biological domain as its point of reference. Autopoiesis, a term that literally means "self-making," began as a biologically based theory of organization in living systems. First theorized by the biologist Humberto Maturana and the cognitive scientist Francisco Varela in the 1970s, autopoiesis-based thinking has since led somewhat of an elusive, yet persistent existence, bringing into play concepts from biology, cybernetics, cognitive science, systems theory, psychology, sociology, and Eastern philosophies.[65] Central to the work of Maturana and Varela on autopoiesis is the idea that cognition is indelibly connected to biological processes. Without suggesting any sort of biological reductivism, they note that our biological structure, our particular modes of sensorially interacting with the world, constitutes a set of cognitive domains, such that each act of "doing" is also an act of "knowing," and vice versa:

> knowing is the action of the knower; it is rooted in the very manner of his living being, in his organization . . . we believe it is necessary to understand how these processes are rooted in the living being as a whole.[66]

In this sense, the "blind spots" of our cognitive modes are structuring principles for how our knowledge of the world is constituted. As Maturana and Varela state, "we do not see that we do not see," and even in our normative modes of cognition, "we do not see the 'space' of the world; we live our field of vision. We do not see the 'colors' of the world; we live our chromatic space."[67] Unlike Bertalanffy, the ISB, and Kauffman, Maturana and Varela raise an issue previously unaddressed: that of the situated position of the observing system. In situating the observer in this way, Maturana and Varela suggest that, on a metalevel, the cognition of cognition, the generation of knowledge is above all a practice, and is in this sense intimately tied to the "biological roots of cognition."[68]

From this methodological perspective, Maturana and Varela outline a theory of biological life that would take into account the doubled character of knowing/doing. In their *Autopoiesis and Cognition,* they address the question of biological life from the perspective of biological "organization." As they state, such an approach is generally based on a systems theory, and would place an emphasis on relations (instead of things), and, more specifically, on relations as indissociable from processes that constitute the living system as a system:

> It is our assumption that there is an organization that is common to all living systems, whichever the nature of their components. Since our subject is this organization, not the particular ways in which it may be realized, we shall not make distinctions between classes or types of living systems.[69]

Again, the subtle addition Maturana and Varela make to Bertalanffy's systems theory is that it is not enough to make an epistemological shift from things to relations. In addressing the question of what makes living systems both "living" and "systems," Maturana and Varela suggest an orientation in which "relations" mean relations that work toward continually realizing the living system as such. In this sense, there is some analogy to Bertalanffy's notion of the "steady state," and to Kauffman's notion of biological self-organization existing at the "edge of chaos."

What are "living systems" for Maturana and Varela? In their dense, often abstract style, they state that a living system is a system defined by its autopoiesis:

> An autopoietic machine is a machine organized (defined as a unity) as a network of processes of production (transformation and destruction) of components that produces the components which: (i) through their interactions and transformations continuously regenerate and realize the network of processes (relations) that produced them; and (ii) constitute it (the machine) as a concrete unity in the space in which they (the components) exist by specifying the topological domain of its realization as such as network.[70]

We can begin to unpack this definition by focusing on some of the key concepts it entails. First, there is the particular way in which Maturana and Varela use the term *machine.* By machine, they do not mean machine technology, nor do they mean anything automated or instrumentalized. By machine, they imply not components or substances (wood, metal, silicon), but rather certain processes and actions (those of "production"); in this sense, the living organism is a machine, simply because it contains within itself a series of processes that continually produce the organism in space and through time. Maturana and Varela later replace the term *machine* with the general usage of the term *system.* The point is, however, that "autopoietic machines" are systems whose processes work toward the continued realization of the system itself.

Second, the phrase "network of processes of production" needs to be understood very specifically. The living system is a network in that it involves more than one process, which itself involves more than one component. In the cell, this is illustrated by the various processes of metabolism (themselves composed of disassembly, assembly, and energy-releasing processes), gene expression (involving structural, regulatory, operator, and other genes), and communication (including processes of cell signaling, membrane transport, protein synthesis). But the networks that compose living systems are not just meta-stable relationships, they are networks of processes, and those processes are predominantly processes of production. The processes of production of living systems imply a set of relationships that is highly dynamic, always changing, but somehow retaining its functional identity, its organization.

The maintenance of the living system as a system through dynamic change (networks of processes of production) is the third part of Maturana and Varela's definition of autopoiesis. As networks of processes of production, living systems (autopoietic machines) have as their result two effects: (1) a production network, which are relationships that make possible the production of the system's self-production; (2) a systemic network, which are relationships that constitute the system, and that undergo perturbations, transformation, and compensations. These two constitutive effects compose the system's organization, for they not only produce the components that continually produce the system (self-maintenance), but they also demarcate a boundary delimiting "system" from "environment" (self-identification). Maturana and Varela also refer to these two effects as a "dynamics" (metabolism) and a "boundary" (membrane).

For instance, there are several overlapping processes and components in the eukaryotic cell. The coexistence of nucleic acids and proteins constitutes the domain of interactions, the boundary of the cell. DNA participates in the specification of RNA and indirectly proteins, just as proteins participate in the specification of polypeptides and other larger molecules. These protein molecules (such as metabolites and lipids) participate in the ordering of the cell's rates of metabolism and regulation of the production of the components.

"Constitution" is done mostly by proteins (regulating protein production via nucleic acids): "constitution occurs at all places where the structure of the components determines physical neighborhood relations (membranes, particles, active site in enzymes)."[71] The mode of this operation is the creation of affordances.

"Specification" is done mostly by nucleic acids (specific process for the production of specific molecules): "specification takes place at all points where its organization determines a specific process (protein synthesis, enzymatic action, selective permeability)."[72] The mode here is an emphasis on processes.

"Order" is achieved mostly by metabolites (regulation of rates of metabolism, replication, expression): "ordering takes place at all points where two or more processes meet (changes of speed or sequence, allosteric effects, competitive and non-competitive inhibition, facilitation, inactivation, etc.)."[73] The mode here is regulation.

A simple way of summarizing this is that in living systems, autopoiesis is the process of the production of the components that produce the system. Autopoiesis is one particular type of systems view of biological life, of organisms as autopoietic machines, as living systems.

In the context of our discussion thus far concerning systems theory and systems biologies, autopoiesis addresses several issues raised by Bertalanffy, the ISB, and Kauffman.

One issue is that of identification. In conceiving of the living system as constituted by an ongoing boundary demarcation, or "act of distinction," Maturana and Varela specify the tension between system and body on the biological level. As we have seen, both Bertalanffy and Kauffman posit a view of living systems that is fundamentally distributed and founded on the notion of relations rather than things. At the same time, there are moments in both authors' texts in which an attempt is made to render as a singular unity that distribution, that network (Bertalanffy through reference to "Man the individual" and Kauffman through the assignation of teleology to self-organization). The response of Maturana and Varela is at once very simple and very complex. A system (a system of networks of processes of production) cannot exist as a system without an act of distinction; that is, the second a relation is articulated, a kind of topology is formed, what Maturana and Varela term an "autopoietic space," in which the relations that make a network a network may be articulated. Despite a network's being distributed, or a system's being "open," there may still exist a dynamic interior and exterior to the network or system. Indeed, as Maturana and Varela propose, such are the conditions of possibility for a living system.

A second issue is that of the relation between machine and organism. The importance of the act of distinction (boundary demarcation) still does not address how a paradoxical identity may be said to exist when the living system not only exists in a dynamic environment, but when it is in fact constituted by dynamic processes internally (the networks of processes of production). Here it is important to note a distinction that Maturana and Varela make in understanding autopoietic systems: the distinction between "organization" and "structure." As they state, a living system may be defined by a single organization, but may be realized in one of many possible structures: "Organization denotes those relations that must exist among the components of a system for it to be a member of a specific class. Structure denotes the components and relations that actually constitute a particular unity and make its organization real . . . Living beings are characterized by their autopoietic organization. They differ from each other in their structure, but they are alike in their organization."[74]

In Maturana and Varela's example, a toilet has an organization of water-level regulation and may be realized by components that are metal or wood, without changing the organization of the toilet. In applying this to the biological domain, Maturana and Varela suggest that it is their autopoietic organization that defines living systems as living (as per their definition given earlier). This also serves for them to iterate the distinction between biological and nonbiological systems, or between organism and machine, for, although machines may be said to perform replication, copying, and re-production, they are generally considered unable to sustain themselves in autopoietic networks of continual, dynamic self-production. It is this capability that Maturana and Varela call "biological autonomy."[75]

Finally, the organization–structure distinction, which serves to also differentiate machine from organism, is a means by which Maturana and Varela address the infor-matic perspective of molecular genetics. Here Maturana and Varela fall somewhere between Bertalanffy, who sees no relation at all between the living system and genetic "codes," and the ISB, for which the integration of systems theory with an informatic view of biological life is not an issue. Specifically, Maturana and Varela state that to discuss genes as containing "information" is an error: that genetic information is trans-mitted ontogenically and phylogenetically through a process of replication. The con-fusion here is between "replication" and "heredity." The latter, as with other genetic processes, is, for Maturana and Varela, a fundamentally transformative and transform-ing process, quite different from the meta-stable replication of information. This con-fusion also grants DNA or genes with an agency they do not have: "when we say that DNA contains what is necessary to specify a living being, we divest these components (part of the autopoietic network) of their interrelation with the rest of the network. It is the network of interactions in its entirety that constitutes and specifies the charac-teristics of a particular cell, and not one of its components."[76] The key to understand-ing the place of informatic views vis-à-vis autopoiesis is that, while traditional molecu-lar genetics views information as stable through replication, for Maturana and Varela, genetic and other types of "information" are defined by both an organization and a structure, which is in itself defined by transformation. A particular genetic organiza-tion, such as transcription, may be conserved through cellular replication; but the materialization of this organization (recall the ISB's "knockout" experiments) may differ widely depending on several factors.

Autopoiesis, in its biological framework, takes up the systems-based approach out-lined by Bertalanffy, but, rather than developing a purely technical response as the ISB does, it outlines a theoretical-biological response. In their formulations, Maturana and Varela raise and attempt to address some of the fundamental tensions in a systems view of biological life. Their notion of autopoiesis as the "organization of the living" is intended to offer an alternative view of biological life, as opposed to traditional molecular genetics, neurobiology, or classical immunology.

"Where is my outline—I start to fade"

Maturana and Varela offer a view to the ability of the living system to constantly main-tain its organization through dynamic internal and external perturbations and com-pensations. Provocatively, they discuss a "biological phenomenology," in which that which the system is capable of withstanding is what defines its mode of biological existence. Although they do not explicitly reference philosophical phenomenology (Husserl, Merleau-Ponty), it is worth considering their remarks in this context: "Consequently, the phenomenology of an autopoietic system is necessarily always commensurate with the deformation that it suffers without loss of identity, and with the deforming ambience in which it lies."[77]

In this sense, biophenomenology is the domain of interactions of which a living (autopoietic) system is capable. Its phenomenology reaches a threshold when its transformation exceeds the system's continued production of its organization. The specific changes that the system is capable of undergoing define its "plasticity."

The notion of a biophenomenology begins to address the difficult philosophical and ethical issues surrounding the tensions between system and body, net and self. Despite autopoiesis's complexification of the understanding of biological life, the question of the "inhuman" remains—the way in which notions of the human are tied to biological presuppositions concerning matter (organic/nonorganic) and form (anthropomorphism). We can put Maturana and Varela's formulation as a question we posed earlier: What is the phenomenology of the biomolecular body? What are the thresholds of plasticity for bodies, subjects, and the self?

Broadly speaking, systems biology, as a systems science, ends up with a series of paradoxes. On the one hand, it emphasizes relations, not things; but, on the other hand, it assumes the ontological difference between organic and nonorganic systems. Why shouldn't a bioinformatics application be considered living, for it retains the same relationships? If we are talking about systems, and displacing the literalism of gene-centric approaches, then doesn't it makes sense to extend systems biology to more than DNA and proteins? Why doesn't it include technical systems, nervous systems, endocrine systems, social systems?

In short, systems biology emerges as a response to the immense amount of data generated by the genome project (what Leroy Hood calls "discovery science"). Its response is a specifically technological—and technophilic—one, relying on the ideology of enabling technologies. In the process, however, it raises a question that is not just technical but philosophical: is the systems viewpoint compatible with notions of the autonomous, closed self? This is where the work of Varela and Kauffman comes in, by addressing experimentally and theoretically the ways in which systems approaches may transform (and challenge) our understanding of biological life.

Systems theory is a conscious response to the dominant paradigm of gene centrism in biotech. In place of the emphasis on genes, it offers an emphasis on interactions between molecules. In place of the emphasis on things, it offers relations. In place of the emphasis on substance, it offers network and pathways of behavior. In place of the use of high tech to target gene candidates, it offers the use of modular, portable, synthetic technology offering a global view. However, in offering this alternative, it does not take into account the larger philosophical issues it opens up, namely, the incommensurability between the self and the network, between body and system. Curiously, it opens onto a nonhuman view of the biomolecular body, one fundamentally at odds with the subject-based, human view of genetic medicine. Alternative theories (Kauffman, Maturana and Varela) offer a mediating point between a nonhuman biomolecular body and a human-subject–based medical genetics.

CONCLUSION

The Bioethics of Metadesign

"Just pretend it's for real"

In the midst of considering technological advance, issues of ethics, accountability, and culpability can easily be lost in the "gee-whiz" rhetoric of the mass-media and pop-science discourse. A form of ethical alarmism tends to pervade this discourse, and too often such questions are raised in the form of science-fictional dystopianism: the "grey goo" problem in nanotechnology, the specter of using human cloning to make a slave race, and the fears of genetic technologies run amok are all instances of such ethical alarmism. To be sure, these anxieties—culturally manifest in science-fiction literature, film, and video games—are expressions of more specific issues. These issues (techno-logical instrumentality, the relationship between therapy and enhancement, and the effect of biotechnologies on what is culturally perceived as "human" and/or "natural") occupy much of the current debates surrounding human therapeutic cloning, research on human embryonic stem cells, and the potential applications of "emerging" sciences such as nanotechnology. We see (or do not see) these more specific issues put into practice at various social and political levels—policy decisions, public and not-so-public debates, conferences and symposia, science education, and modes of activism.

We can begin by saying that there is a trajectory from the most fanciful and exagger-ated dystopia in science fiction (films such as the *X-Men* series or *28 Days Later*) to the most specific analysis of a particular issue in biotechnology (in the deliberations of the President's Council on Bioethics over the resourcing of human embryonic stem cells).[1] Often the point of mediation from one end of the spectrum to the other is the way in which the human subject is positioned vis-à-vis biotechnologies. In the "official" contexts of bioethics in the United States, individual and collective agency, technolog-ical democratization, and the empowering potentials of technological development are often taken to be primary values. These values are, of course, tied to various locales, be they institutional (NIH-funded research at a university), governmental (national policy decisions regarding biotechnology), or corporate (proprietary technologies and restricted access to information).

At the root of these value systems are sets of assumptions concerning fully agential human subjects who make use of emerging technologies as tools for specific human ends. What follows here is an experiment that questions such assumptions. The questions raised here are difficult, frustrating, and often without apparent relevance. But the argument made is that, alongside current applied bioethics practices, we would do well to consider a parallel "bio-ethics" that would critically investigate the assumptions informing our national committees, policy debates, and activist principles.

The reasoning behind this argument is the following: If emerging biotechnologies are transforming our basic notions of "body," "nature," and the "human," then it follows that a bioethics capable of accommodating such transformations would itself have to be adaptive and flexible. And this might very well entail a critique of the philosophical foundations of applied bioethics. Although we can certainly apply rigid categories of the liberal-humanist subject to biotechnologies, in the context of patient-specific biochips, on-line genome databases, in vitro tissue engineering, and DNA computing, such categories seem to obfuscate more than they reveal. Where is the subject in the application of genetically specific biochips, which provide disease predisposition analysis for a patient by accessing a genome database, which then returns results designating possible pharmacogenomic treatments for preventive genetic medicine? In one sense, the answer is quite easy: the subject is the patient. But where exactly is "the patient"? Is the patient the anatomical-physical body that offers a vein for a DNA sample? Is the patient equal to the DNA sample? Is the patient equal to its computerized genetic profile? Is the patient the one who gives blood, or who takes in a drug? It seems that we do not yet have a language or body of concepts for describing such instances.

In this sense, the work of science studies has been helpful in developing new ways of talking about our ceaseless interactions of bodies and technologies. Bruno Latour's notion of "actants" is one such example. In his study of the role of technical mediation in Pasteur's germ experiments, Latour shows how the reframing of the site of experiment as a laboratory involves mobilizing a number of people and things into specific relationships.[2] The "natural" field of the farmland is transformed by Pasteur into a laboratory, in which the observation of animal physiology and germ transmission is foregrounded. In this laboratory, Pasteur acts as the one who "delegates" partial agencies of an array of things, which fulfill their role of objective laboratory measuring instruments. The laboratory instruments are not agential subjects in the same way human subjects are, but by having certain partial agencies delegated to them, they participate in the human activities of, in this case, scientific experiment. Oftentimes we take little notice of this mode of delegation, where mere objects become "actants" (as opposed to actors). When we do notice it, it is often when the "black box" breaks down, eliciting a technical frustration from us human "users." When the actant breaks, it can no longer fulfill is delegated agency, and the whole activity—actors and actants combined—similarly comes to a halt.[3]

Concepts such as "actants" elicit something paradoxical: a nonhuman bioethics, or, as Gilles Deleuze states, an "ethics without morality." Before moving to such claims, however, it will be useful to attempt to characterize contemporary bioethics, paying particular attention to its philosophical foundations.

Bioethics

Contemporary thinking about bioethics is roughly divided between two approaches. The first is a "philosophical ethics" applied to issues in biomedicine, or better, the extraction of philosophical ethical issues from particular biomedical examples (e.g., in vitro fertilization [IVF] or stem cells and the definition of "potential life").[4] The second approach is "applied bioethics," and draws on sociology, anthropology, and practical ethics to analyze issues related to specific contexts (e.g., risk assessment in genetic screening as part of genetic counseling for parents).[5] Broadly speaking, the trend in bioethics today—and mostly in the United States—is toward the latter approach, which assumes the theoretical issues set forth by philosophical ethics (defining "life," "freedom," "choice," etc.), and which performs analytic and prescriptive studies based on case studies, fieldwork, and/or analyses of observations or data. A common example of applied bioethics is the studies of risk assessment in the handling of genetic information, or the studies of informed consent in cases of biological patenting, or the studies of ethical practices in scientific experimentation with human embryonic stem cells.[6]

To these two approaches—philosophical ethics and applied bioethics—we might add a third approach, that of a "cultural bioethics."[7] Bringing together approaches and concerns from cultural studies, poststructuralism and postmodernism, and science and technology studies, the cultural bioethics point of view stresses the mediations of scientific knowledge within bioethical debates and discourses. Inasmuch as all cultural theory concerned with the politics of subjectivity implies an ethics, the cultural bioethics standpoint engages with ethical issues on a much more flexible, theoretically broad level. Some approaches focus on the ways in which scientific knowledge is disseminated in popular culture (in films, news media, TV), while other approaches are concerned with how the intersection of biology and technology influences cultural notions of subjectivity (in the workplace, as part of consumerism, in medical technologies). Still other approaches situate science and technology in a social and/or institutional context, examining the ways in which knowledge is produced, practices developed, and artifacts mobilized as points of mediation (in the laboratory, in the hospital, in computer databases).

Although the cultural bioethics approach is generally marginalized in the mainstream bioethics discourse (at the level of medical education, bioethics organizations, policy committees), its concerns overlap both the philosophical ethics approach and the applied bioethics approach. From one standpoint, it may be said that cultural bioethics, moving as it does between the specialist and nonspecialist, science and science fiction, the theoretical and the practical, serves as a mediation between the universal interests of philosophy and the specific interests of applied ethics. However, at least one point of tension that arises here is that, by and large, cultural bioethics is predominantly concerned with critique as its main mode of activity. This is different from philosophy, whose main goal is an articulation of general concepts (such as the development of ethical principles), and it is also different from applied ethics, whose main concern is the pragmatics of specific medical-ethical situations (such as genetic privacy rights in health insurance).

Critique, as Michel Foucault has pointed out, works at the interstices of its object, revealing the points of fissure in the forces that come together to form a given practice, discipline, a given body.[8] Critique is not merely the "negative" work that is necessarily done, so that a subsequently "positive" resolution may follow. Critique is a generative practice at precisely the moment of its negativity; it therefore provides openings, pathways, and alternatives that were previously foreclosed by the structure of a discourse. Critique in this sense is not purely philosophical, for it is less concerned with defining or creating concepts than with generating what Foucault calls "problematizations." It is also not applied ethics, for its point of interrogation lies not at the level of problem solving, but at the infrastructural level of assumptions and foundations. From this point of view, the critical aspect of cultural bioethics is frustrating: it appears to offer no help whatsoever, only analyses that point to more problems, challenges, inconsistencies, and fissures—not the best antidote for the instituting of policies, or the manifestation of laws.

However, this approach is significant if only because it points to the fact that often the challenges that bioethics faces have to do not with discrepancies in the details of ethical policies, but with more fundamental, philosophical questions. At first it may seem strange to think of bioethics—predominantly concerned with questions such as whether or not the U.S. government should outlaw human cloning—as a philosophical enterprise. The main bioethical issues of consent, brain death, abortion, and IVF all pose fundamental questions that are philosophical in nature, and in this way each issue is simultaneously philosophical and imminently pragmatic. Bioethics' history can be seen as a series of contexts in which specific bodies (patient bodies, physician bodies, viral bodies, chemical bodies, cellular bodies, populations, species bodies) are re-formed through discrete practices (injection of therapeutic drugs to counteract the symptoms of a genetically based disorder). The bodies, though never apolitical, become politically materialized at the moment they are transmuted into policies, laws, governmental guidelines, funding sources, marketable and FDA-approved drugs, and medical-economic investments and insurances. A few brief examples will make this clear.

Subject/Object. The dark past of eugenic experiments on human subjects raises the pragmatic question of "informed consent," especially in relation to patentable biological materials from nature or human beings.[9] This has necessitated guidelines for human-subjects research, as well as guidelines for what qualifies as informed consent. The philosophical issues consent is tied to have to do with epistemology and the relation between bodies and knowledges. Is the body an object? Are there instances when it is feasible to treat the body as a "thing"? How is the body as a subject (the embodiedness of the patient) going to be made commensurable with the body as an object (an object of study, of experiment, of diagnosis, of the "medical gaze")? Here philosophical inquiries into materiality rapidly fold onto the economics of materialism, in the commodification of biological entities.

Death. The development of medical and surgical diagnostic technologies since the 1960s (MRI, CT), as well as the various life-support technologies improved by micro-electronics and computer technology (heart–lung machine, dialysis, pacemaker), made it possible to render the moment of "death" as a variable.[10] Prosthetic organs enabled what was once a fatal organ failure to be supplanted by viable nonorganic substitutes; medical visualization made possible a refined diagnosis for diseases such as cancer tumor detection; and life-support machines enabled patients suffering trauma to maintain a minimum level of biological sustenance rather than dying. These technologies produced the medical cyborg, the virtual patient, the living cadaver. They have raised pragmatic issues concerning technological sustainability in artificial organs, the need for responses of high transplant demand, and the more technical definition of "brain death" and the possibility of euthanasia. Is death a technical variable? Is death something that, in medicine, needs to be situationally defined, adapting to particular contexts (vegetative states) or new technologies (life support)?

Life. The question of death in life-support technologies is accompanied by the question of life with reproduction and abortion issues. Pharmacotechnologies introduced during the 1960s such as the pill, legal decisions surrounding abortion (*Roe v. Wade*; RU-486), and IVF have all similarly modified the boundaries separating natural and unnatural conception. The pragmatic issues they raise have to do with the debates surrounding abortion in relation to screening techniques (preimplantation diagnostics), as well as issues pertaining to IVF itself, surrogate mothering, the existence of sperm and egg banks, and the resourcing of biological materials for science research (stem cells). Philosophically what is at issue is how we define life. Like the issues surrounding the definition of death, the question of life and the new reproductive technologies pose broad debates about the role of science in making non- or extra-scientific judgments. The debates surrounding abortion (at what point is there "life"?) are carried over in IVF/NRTs (New Reproductive Technologies), but they are now also infused with the rhetoric of genetics and biotechnology.[11]

Body. More recently, the rise of an informatic biotech industry has made the issue of economic interest in relation to information more urgent. Who controls the data, who has access to the data, and even how the data is defined (where it comes from, how it is processed, how it will be used or misused) are all key concerns in a new bioethical branch associated with genomics, bioinformatics, and pharmcogenomics, and their relation to the economic interests in biotechnology. The pragmatic issues have to do with genetic privacy and its relation to the workplace, health insurance, and medical practice.[12] The broader philosophical questions have to do with how "value" is defined, and what the relation of value is to the body of the subject.

To summarize, despite its highly pragmatic character, bioethics (especially applied bioethics) always implies two concurrent strands that inform its pragmatics: a philosophical strand, in which particular issues are often carriers for more complex, and indeed often irresolvable questions, and a technical strand, through which bioethics issues are contextualized and reframed. Bioethics asks: What is the point of contention, what is the problem? Who are the parties involved? How can policies, laws, and guidelines be developed to effectively respond to the problem? A working assumption here is that bioethics deals with the conflicting agencies of human subjects using tools and techniques in the fields of medical practice and bioscience research, and that the best way to respond to these conflicts is to develop broadly applicable rules that would hypothetically cover every instance.

From the perspective of critique, we can ask seemingly naive questions here: Why only subjects, and why only human subjects? Are there nonhumans involved, are there relationships other than agency involved? Why general rules? Why one set of rules for every instance? If the output of applied bioethics is a rule, is this not a continuation of the "categorical imperative"? If ethics is a practice, what is the role of unaccountable contingencies? What, indeed, is "practice" and how is it different from a rule or guideline?

Where does a traditional ethical approach (as we see in applied bioethics) begin from? We can take Kant's ethical writings, in particular the exposition of the "categorical imperative," as a starting point. This ethical philosophy has, perhaps more than any other, influenced a great deal of the ethical and bioethical debates. At both the level of discourse ("When is it feasible to do this?") and the level of policy ("How can ethical principles be instantiated as policies?"), the traditional background of bioethics is related to the notion of a general law. According to Kant, the "ethical" component of the categorical imperative resides not in subjectivity, not in embodiment, but rather in the capacity of reason to be actuated in a will, which is expressed in the "ought" of the imperative. It can be said that the varying ethical schools—deontological, axiological, situational, noncognitive, utilitarian—all stem from this basic premise of the actuation of a law, expressed through reason.[13]

In the *Foundations for a Metaphysics of Morals*, Kant sets out to establish the grounds for a universally valid ethics based on "unconditional rational necessity."[14] A rational necessity is those requirements that must be fulfilled in order that any action or behavior be considered as rational. It is rational because it is action governed by reason, and not by appetite, instinct, or desire. A conditional rational necessity involves a means to an end (in order to obtain goal A, you must do 1, 2, 3, etc.). This is a "hypothetical imperative," which sets a goal and in setting a goal also prescribes actions geared toward the realization of the goal. By contrast, an unconditional rational necessity involves actions that have no other end than themselves (no matter what, you must do 1, 2, 3, etc.). They are not means to an end; they are ends in themselves. They express the necessity of acting not out of self-interest or goal-orientedness, but out of obligation and duty. For Kant, they constitute "moral principles" that lead to the "categorical imperative":

> All imperatives are expressed by an ought and thereby indicate the relation of an objective law of reason to a will that is not necessarily determined by this law because of its subjective constitution ... there is one imperative which immediately commands a certain conduct without having as its condition any other purpose to be attained by it. This imperative is categorical. It is not concerned with the matter of the action and its intended result, but rather with the form of the action and the principle from which it follows; what is essentially good in the action consists in the mental disposition, let the consequences be what they may. This imperative may be called that of morality.[15]

In considering what might count as moral action, Kant makes a distinction between a subject acting from "good will" and a subject acting out of obligation or duty. For Kant, acting out of a "natural inclination" to good and acting from "duty" are two separate things; the former cannot count as being governed by reason, only the latter can. A natural inclination can serve both a common good and self-interest, but it will always operate from a purpose. An act of duty, by contrast, sets upon the subject an external obligation, a requirement that is it necessary that one act in such and such a way.

A second assumption Kant interrogates is the notion that reason in action is driven by the "content" of the action. However, the concept of duty/obligation in unconditional rational necessity presupposes any specific instance; it prescribes actions whose contents are not specified. Many actions can, depending on the instance, fulfill the same unconditional rational necessity. The emphasis is not on the action, but on the presence of a law; not on the content (specific instance), but on the form (the action in itself). In this way, Kant suggests that moral action, in order to be considered valid, must therefore proceed from pure reason, that is, from a notion of prescribed moral principles that operate above and beyond any one specific instance. They are, in this way, a priori.

Both of these points are captured in Kant's Universal Law: "there is only one categorical imperative and it is this: Act only according to that maxim whereby you can at the same time will that it should become a universal law."[16] Moral action in Kant is therefore adopting those maxims that you would will to become laws. The Kantian maxim implies the enacting of two types of speculations with regard to moral actions: putting oneself in the position of being acted upon (individual aspect), and positing that everyone act in such a way (collective aspect).[17]

The main issue with the categorical imperative is freedom and morality in relation to the subject. Philosophically speaking, if Kant can show the will as free, he can also demonstrate that the will is rational (and self-determining). This requires the formulation of a morality that is universal and objective (not bound to any one situation, and valid across situations to individuals generally). Moral laws must therefore have "universal necessity." It is for this reason that Kant suggests that morality cannot be based on pleasure, because this is constrained by individual situations (they can only be verified experientially). Any maxim describing particular actions can only attain the status of universal necessity (as a moral law) according to its "form" and not its specific "content." Moral laws can be tested by applying them as universal laws, according to their form. This removes the will from the specificity of the phenomenal domain and places it in the space of speculative reason itself. This speculative aspect of practical reason serves as the indicator of the potential of which the individual is capable ("ought" implies "can").

Kant's analyses up to this point privilege forms of legislation combined with forms of self-determination into a vision of the liberal-humanist subject (a moral site he calls the "Kingdom of Ends"). The truly free and moral subject is one who acts "autonomously" (acting governed by the moral law) rather than "heteronomously" (acting governed by other interests or rewards). Therefore, the categorical imperative is in many ways equivalent to the law of autonomy. The Kingdom of Ends is Kant's utopian moral community in which laws are collectively determined, and the laws respect individual autonomy and self-determination. Interests are shared and go toward supporting autonomy. There is, however, a tension in this logic: to be truly free, we must consent to laws, and rational self-determination is only legitimized by the establishment of metahuman, a priori concepts. We cannot be motivated by the moral law unless we are free, but we cannot be free unless motivated by the moral law.[18]

There are a number of comments in Kant's ethical philosophy that seem second nature to us; there are also a number of propositions that seem unethical or philosophically vague. The point in raising the Kantian categorical imperative is simply to suggest that in critically analyzing contemporary bioethics, we should consider the ways in which ethical foundations may inform how bioethics asks its questions, how it approaches a problem, and the concepts it assumes to be universally valid. Through a Kantian lens, we can see how a traditional ethical approach begins with a methodological goal: the formation of principles that may be materialized as law (in both ideology and policy). An example are the debates over research on human embryonic stem cells. One of the primary issues in this debate is whether stem cell research is an instance of medical therapy or extramedical enhancement. Much discussion has to do with whether the U.S. government should continue funding such research, knowing that one of the resources of stem cells is discarded embryos from fertility clinics (artificially inseminated eggs that were not used in IVF procedures, and that would be destroyed anyway). However, stem cells (and there are many different kinds of stem cells) exist in many different kinds of contexts, and it is unlikely that a single set of guidelines will be acceptable across all possible uses of stem cells in research. Behind these ideological, ethical, and economic deliberations is a more troubling question:

that, with developing biotechnologies, the very notions of what counts as normative health may be in the process of being redefined.

If much current bioethical activity in some way assumes Kantian ethics (specific instances, universal laws), then there are a number of issues that such approaches elicit.

Ethics, bodies. The basis for a bioethics is, to state the obvious, an articulated relation of the social to the body. "Body" here means several things: the body as articulated and defined by science (biology, anatomy, genetics), and the body as acted upon, as treated, by science and medicine. In addition, the body is taken here both as proper to an individual subject and as a collective body of citizens. This latter description—individual and collective—is part of Foucault's notion of "biopolitics": the practices of regulating the social—the population—defined as a collective biological entity (birthrates, infant mortality, demographics, health records).[19] Therefore the body is both the subject and object of bioethical considerations: both the question of what is good for the body (my body, your body, the body of society, the bodies of people), and the question of how the body can be good (what techniques, what technologies, what practices, what policies, what services, what therapies).

Bodies, subjects. If the body were an inanimate object, bioethics would become contradictory; the body would then be a resource, a thing, a commodity. What makes bioethics ethical, and not just "bio," is an assumption that the self is intimately related to the body. Although the exact relation between self and body widely varies, bioethics is founded on the notion that the body is also a subject. Whether the subject derives from the body (sociobiology), or whether the body is formed by the subject (liberal humanism), is not the issue here; rather, it is the assumption that, although the residues of Cartesian dualism may still be present, the body is the subject, at the same time that the body is different from the subject.

Subjects, ethics. Bodies do not act in a vacuum, and neither do subjects. The varying forces of government, economics, institutions, and culture all permeate the body at the most micrological levels. Therefore, bioethics is not just concerned with the good in relation to bodies and subjects, but, in its interest in the ethical and moral dimensions of bodies, it is at the same time interested in the ways in which notions of good and bad are manifested in bodies. Acting and reacting, bodies profusely materialize ethics and morality of the culture and society in which they act and react. This relationship too is filled with differing arguments; Kant begins from an a priori categorical imperative, whereas contemporary cultural theorists argue that there is no body that precedes ethical action.[20]

Ethics, laws. The passage is from the universal law described by Kant to the formation of governmental policies as the primary way in which bioethical judgments are established. It is almost humorous to think of ad hoc, informal, improvised ethical protocols as a course of policy (and yet these are part of the daily ethical practices in medicine, for better or worse). Bioethics, in its governmental guise, is predominantly a top-down affair, with scientific knowledge and bioethical issues largely mediated by the mass media, and the final word on issues demonstrated by the negotiations over policy. However, this is only part of the picture. Another, perhaps more important, source of legitimation in bioethics is within the biotech industry, where a different kind of juridical process takes place. More flexible than governmental laws, consumer biotech offers a range of perspectives on ethical and social issues, from funded art exhibits and documentaries, to expanded pharmaceutical markets for consumer medicine, to the establishing of in-house ethical offices (often connected to public relations departments). Leading bioethicists such as Peter Singer have noted the general shift in bioethical concerns away from governmentality, toward consumerism; whether this is the case remains to be seen (especially in light of many academic–corporate partnerships), but what remains worth noting is that the legitimacy of the ethical body is its materialization in some form of (juridical or market-driven) law. Here bioethics is to be found in the President's Council on Bioethics, the NIH, and Celera Genomics Corporation.

Although the Kantian ethical paradigm raises philosophical issues, its combination of empty formal maxims and universal laws makes for rather rigid ethical protocols. And, any given issue in biotechnology is much more complex than a single policy for all cases. While not denouncing the Kantian paradigm in total, and while not devaluing the ongoing work in practical ethics, we might also inquire into other paradigms for bioethics.

Bio-ethics

The first thing to note is that "bio-ethics" is quite different from "bioethics." Whereas bioethics describes a discipline with particular modes of practical application, bio-ethics is by comparison a vague, highly relational, and critical-philosophical mode of inquiry. Whereas bioethics aims to arrive at near-term resolutions to highly specific challenges (e.g., guidelines for research on stem cells), bio-ethics outlines a long-term polyvalent "opening" of possibilities for ethical thinking (e.g., what is a "nonhuman bioethics"?). Whereas bioethics addresses problems that are the culmination of science, technology, and medical practice, bio-ethics takes up the potential implications of design relationships between bodies and technologies. If bioethics is concerned in a utilitarian manner with contemporary issues such as human cloning, stem cells, NRTs, genetic patenting, and genetic privacy, then bio-ethics is experimentally concerned with what "bio" means in relation to "ethics," with the ways in which ethics always implies a "bio" component. Bioethics shows us the congruency of practice with principle; bio-ethics critically evaluates our own limitations in thinking bioethically.

The reason for stressing the hyphenated term *bio-ethics* is to take literally (that is, in its materialization) the notion of an ethics related to and informed by bodies. "Bodies" can mean many different things, and in more traditional bioethics, bodies usually refer to anthropomorphic, medical, and genetic bodies. The body, as a starting point, is one of the main points of diversion in distinguishing bioethics from bio-ethics. If the body in bioethics is medical-anatomico-genetic, the body in bio-ethics is defined by parameters that are qualitatively different, while never denying the contingencies of medical-anatomico-genetic norms.

The orientation of bio-ethics can be made clearer by a philosophical shift. If bioethics takes up the Kantian project of categorical imperatives (moral laws), then bio-ethics takes up the project begun in Spinoza's treatise *The Ethics*.[21] Spinoza's *Ethics* offers an instructive counterpoint to Kant's *Metaphysics of Morals*; although both are concerned with how valid modes of ethical action can be made possible, they widely differ in their terms and concepts. Deleuze, writing about Spinoza, defines the body along two main axes and an initial proposition.[22] The proposition is that a "body" is any set of particles that stand in relationship to each other, such that the relationship is one of composition (rather than decomposition).[23] "Bodies are distinguished from one another in respect of motion and rest, quickness and slowness, and not in respect of substance."[24] A body thus articulates aggregates of particles that may have a molar effect of a single unit (a whole cell, a whole organ, a whole anatomy, a whole self, a whole couple, a whole community, etc.). Indeed, in this sense, a body can be any such aggregate (a cell, a society, a set of concepts, an interface, a network).

Taking the rapidly changing fields of biotechnology such as genomics, proteomics, and regenerative medicine, we can ask the question Deleuze's Spinoza asks: "what can a body do?"[25] In the context of biotechnology, this question is threefold.

What is a body? Deleuze's Spinoza always comments that we do not know what a body is, or what a body can do. This is never more true than in the technosciences, where new developments daily redefine the boundaries of the body/technology, natural/artificial, subject/object. The seeming empiricism of this question has embedded in it a deeper question concerning the ways in which "the human" is variously defined. Knowing what a body is, and how a body is related to a person, means that ethics becomes much more than the prescription of protocols; it becomes an inquiry into the meaning of the human.

What can a body do? How is this question asked in genomics? How is it asked differently in personalized health care? Knowing what a body can do means knowing what a body is in terms of ethical considerations. It also implies that a knowledge of the body is coextensive with a potential instrumentalization of that body.

What can you do to a body? How is this question framed by genetic engineering, by the debate over what counts as "enhancement" in biotech? Does this question always imply instrumentality? Knowing what you can do to a body means being able to define a flexible, adaptive ethics that is sensitive to the historical developments and technical contingencies of biotech. Knowing what you can do to a body also means understanding the body as both capable of complex affects and, for this reason, uniquely vulnerable.

With this proposition, bodies are defined first by their kinetic modes: relations of motion and rest, relations of speed and slowness. The particles of a "body" are static or changing, speeding up or slowing down, juxtaposed and cumulated. Even when asleep, our bodies are ceaselessly active, tossing and turning, heating up and cooling down, bodies in dreams and bodies that awake from dreams. Bodies are also defined by dynamic modes: the capacity to affect other bodies, and the capacity to be affected by other bodies. There are always other bodies, be they things, animals, or people, and there are always degrees of interaction with other bodies. The more complex a body, the greater its affective component. "Affect" in this sense is more than mere "feeling" or emotion; affect is a differential in force accommodated by the mode of a body at a given moment—what a body is capable of. The important thing is not feeling, and not being able to feel (being sensitive); nor is it sympathy (feeling for) or empathy (feeling as if). The important thing is the indistinguishability between feeling and self, and the way that both feeling and self are constituted through and through by modes of individuation, or what Deleuze calls "nonsubjective affects."

In short, both bioethics and bio-ethics have to do with the body, but each defines "body" in a different way. If bioethics assumes a view of the body defined by the traditional sciences of medical anatomy and genetics (a constructivist, mechanistic body), then bio-ethics begins by questioning the body (motion and rest, affects). This means, in turn, that "ethics" has quite a different meaning for each approach. For bioethics, ethics is inextricably linked to morality, to moral law. If bodies are discrete, quantified objects (the medicalized, patient body, the anatomical body, the body constituted by its parts, by DNA), then an ethics will be one in which treatments, manipulations, and controls of those object bodies will be prescribed. Bioethics will mean "how to handle bodies" understood as bodies "standing in" for subjects. Following the Kantian paradigm of the "categorical imperative," moral law informs and makes possible ethical practice. Knowing what can be done to a body would thus imply its moral echo, what should or should not be done to a body. Again, "body" is here assumed to be consonant with the human subject, just as the body defined in the medical-anatomico-genetic paradigms maintains some essential relationship to the "person" occupying that body.

In bio-ethics, we have what Deleuze calls an "ethics without morality," a differentiation between ethics as an immanent practice and morality as the instantiation of a law.[26] If, in Deleuze's Spinozism, bodies are primarily defined according to their affects (their capacity to affect, their capacity to be affected), then we make two shifts from the bioethical position: First, we shift from approaching bodies as objects to approaching bodies as relations, relations that are equally extensive (materialized) and intensive (contextualized). Second, we shift from assuming bodies to be congruent with subjects, anatomies with persons, and move toward considering bodies as "relations of composition" or "relations of decomposition." Bodies are certainly connected to matter as such, but instead of discrete "things," they are found to be articulated through motion and rest, speed and slowness of particles. These bodies, indissociable from "what a body can do," constitute for Deleuze a kind of "ethology," an account of how bodies affect other bodies, and how bodies are affected by other bodies.

At times, the relations of composition may be considered "good," in that they compound the active forces between bodies.[27] At other times, the relations of decomposition may occur, a "bad" relationship in which bodies stand in relations of decomposition to each other. This bad relationship can be terminal and detrimental, as in the case of poisons (bodies of chemicals decomposing the relationships of the bodies of blood cells, which are connected to bodies of the respiratory system). In extreme cases this results in such a radical transformation of relationships (that is, decomposition happens to such an extent) that the bodies that result (the relations that result) are incommensurate with the relations with which we began. This is not just death, but, as Deleuze's Spinoza points out, a whole new set of relationships acting in a relation of decomposition to bodies. However, the "bad" is not always "evil." Bad relationships, in part or in moderation, can also be agents of transformation, as in the use of vitamins, homeopathics, herbal medicines, or even the regulation of neural firing patterns in the brain.

Instead of assuming that bioethics has to do with the good relationship between bodies and medicine, Spinoza does not assume that we understand the full implications of bodies. Spinoza's question of what a body can do is accompanied by an analysis (via Deleuze) that takes bodies as kinetic and dynamic. Before assuming that we know what a body is, Deleuze's Spinoza suggests that bodies are always changing, and that they are always interacting with other bodies. These are two observations that would seem highly relevant to bioethics.

This consideration of what a body can do also means that "body" be taken not as a universal, but as highly contingent: sociohistorically and scientifically contextualized by genetic histories, genome therapeutics, regenerative medicine, biological patents, telemedicine, and health-care patient databases. The question of what a body can do therefore opens onto the question of the ways in which the body is articulated, materialized, articulating, materializing. Institutions (hospitals, medical schools), technologies (MRI, genome databases, DNA chips), modes of governmentality (distribution of federal funds for research, genetic privacy), and modes of consumerism (pharmacogenomics, outpatient monitoring) all play a role in this seemingly abstract, simple question.

This means that when beginning from a bio-ethical point of view, the philosophical question of what a body can do is simultaneously an ethico-political and technoscientific question. In the case of genetic diagnostics, for example, what a body can do is delimited by DNA chips (a body can "express" itself genetically in relation to a normative database); laboratory technologies (a body can remain biomolecular, genotypic, only reaching the phenotypic patient at the stage of reportage of results); genome databases (a body can serve as both standard and patient; a body can be information); genetic drugs/therapies (a body can be repaired not by treating symptoms but by modifying molecular code); genetic screening processes (a body can be a time machine, a garden of forking paths; a body can be accounted for by its own data as insurance for the future).

The traditional ethics approach in bioethics gauges the effects of science and technology, of medical practice, against a set of ambiguous values related to "the human." Oftentimes, widely diverging practices and technologies (such as IVF and stem cells) are boiled down to a single value set (inherent value of life versus life as a commodity). The debates at the governmental level are both embarrassing and illuminating, for they inadvertently demonstrate the inconsistencies in the approach of traditional ethics and the law (e.g., the U.S. government's concern over life-as-commodity covers the main issue, which is how the government should spend taxpayers' money) (Figure 20).

The central tension in the bio-ethics proposed by Deleuze is that it requires us to forgo some deeply entrenched modes of thinking and acting. First, it calls on us to think of the possibility of a nonhuman bioethics; that is, an ethics that, while it places human concerns central to itself, also does not prescribe human interests, as if the human could be effectively separated from a milieu in which specific embodied humans exist. Second, it suggests that ethics is not a program, a prescription, a preset law or rule, but that it is a pragmatics, a practice, a modality, a state of relations between bodies. Third, it points to a fundamental tension in current bioethical thinking: that the model for ethics is that of the juridical-Cartesian model of the subject (accountability, agency, culpability, instrumentality). This legal instituting of the subject is the main block to fully realizing a bioethical ethology. It is implied as early as the Hippocratic oath, where certain subjects are responsible for certain actions under a code or law.

For Deleuze's Spinoza, bodies are relations, and subjects are affects, both a geometry of shifting states, of modes of attributes of a single substance. Ethics equals affect for Deleuze's Spinoza. Rather than take up a discourse at its given ends, or its nodes ("is it moral to clone human beings?"), Deleuze's Spinoza takes up ethics at its edges ("don't morals and bodies mutually exclude each other?"). Ethics is bodies. What are bodies? They are modalities, that is, relations of states of change. As modalities, they are constituted by two main dynamics: speed/slowness and capacity to affect/capacity to be affected. The more complex the body, the more complex the affects. Ethics is not about doing this or doing that to others (doctor to patient, patient to self); ethics is about the bodily modalities that define an "event." What is the capacity for bodies to relate to each other, to affect each other and to be affected by each other? That is the "ethology" in Spinoza's ethics of which Deleuze speaks.

Is there a morality in Deleuze's Spinoza? Yes and no. There is, certainly, a consideration of values in the relations of bodies. Relations of composition are "good" and relations of decomposition are "bad"; those that increase the body's capacities to be active, affective, are good, those that decrease them are bad. But this good and bad is not

Kantian/Bioethical	Deleuzian-Spinozist/bio-ethical
The individual	Modes of individuation
"Body" (medical-anatomical; physical; mechanistic; anthropomorphic)	"Body" (relations of motion/rest; capacity to affect/to be affected)
Subject/object (mind/body); juridical model (accountability; rational decisioning)	Parallelism (power relations); affective model (embodied interactionisms; "know-how")
Ethics = morality	Ethics versus morality
Good/evil	Good/bad
Moral law (categorical imperative)	Ethology (modes of existence)
Universal, necessary, causal (Kant)	Immanent, relational, contingent (Spinoza)
Security, prevention, negativity (military-juridical)	Flexibility, adaptation, activity (communitarian)
Individual versus society (investment)	Affection and "other" bodies (divestment)
Laws, policy, protocols, principles	Practices, knowledges, guidelines, "plan"
Values	Modalities
State/nation (governmentality)	Multiplicities/collectivities (critique)
Human-centered	Nonhuman-oriented
Design-as-instrumental (engineering before design; ethics as afterthought)	Design-as-ethical (ethics as immanent in design and engineering)
How can ethics be prescribed? (the law)	How can ethics become immanent? (practice)
What qualifies as ethical action? Categories.	Where does ethics occur? Event.

Figure 20. Two bioethics paradigms.

good and evil.[28] Deleuze's Spinoza therefore writes of an ethics without morality; there are no universal laws, no overarching guidelines for action, no empty formalisms, contra Kant's emphasis on the form of the law rather than its content.

Thus, to the bioethics which characterizes ethical standpoints toward the uses and meanings of biomedicine and biotechnology we can contrast a bio-ethics that undertakes a more philosophical, more experimental approach to critically examining the current assumptions and future possibilities for a socially and politically meaningful ethics of transformation. The former—bioethics—is aligned with viewpoints that are both formal and informal, both legitimized and nonlegitimized voices. It is both a discipline (as in applied bioethics) and the contestation of a discipline (as in activism aimed at biotech research). In this sense, both "official" groups, such as the President's Council on Bioethics, and unofficial groups, such as bioactivist organizations, are taken as bioethics. They share the commonality of applying moral principles to a range of biomedical and biotechnological contexts; morality informs ethics, even provides the basis for its existence. In other words, bioethics—official or unofficial, conservative or radical—involves some engagement with "moral law" as elaborated by Kant. The debates surrounding current issues such as stem cell research or human therapeutic cloning revolve around whether its application may have effects that contain a social "good," or whether their implications contain the seeds of a social "evil." Although conservative groups such as the President's Council may stand opposite bioactivist groups, and although the economic interests of corporations may oppose both, the common understanding between them all is that bioethics is fundamentally a moral issue, a matter of deliberating good and evil.

Bio-ethics is something different, and something more difficult. Deleuze's elaboration of Spinoza imagines an "ethics without morality," an ethics that is an "ethology," a "typology of immanent modes of existence," a practice of "nonsubjective affects."[29] What does all this rhetoric of dynamics, kinetics, and affection mean in the context of ethics? Indeed, given the way that Spinoza defines bodies and subjects, why is Spinoza's work titled "*The Ethics*"? It seems paradoxical, or at least self-contradictory, to postulate an ethics without morality. This is not surprising in that philosophical ethics in the Western tradition has more often than not aligned ethics with morality.[30] Even everyday notions of ethics are difficult to distinguish from morality, with definitions circling upon each other ("ethics: having to do with right and wrong, with what is moral"; "morality: having to do with right or wrong, with ethical behavior"; and so forth).

But we might again rephrase Deleuze's question: does ethics necessarily imply morality? If ethics is a practice, and morality the law that informs the practice, then all we have done is to reiterate Kant's distinction between moral prerogative and moral law (where the moral law is a form, irrespective of specific instance or content). If morality is the principles that makes ethical thinking possible, then again we echo Kant's configuration (moral principles, when taken as laws, enable ethical action that is both universal and necessary).

To dissociate ethics from morality seems to tug at something much more fundamental than a reliance of prescribed sets of rules for acting; it points to an infrastructure in which it is possible to assign and assume values. These values pertain specifically to human subjects and their modes of being in the world. Even when they relate to nonhuman entities (animals, plant life, crops, ecosystems, the environment), their frame of reference is the human subject and how it affects that nonhuman entity. It almost goes without saying that ethics is a human endeavor, that when we discuss right and wrong, good and evil, we are in effect discussing a set of human values for human action.

In this way, bio-ethics is not to be opposed to bioethics. Bio-ethics is not simply on the side of "ethics" while bioethics is on the side of "morality." There is a theoretical and a practical point to be made here. On a theoretical note, a bio-ethics is not devoid of any moment of valuation, of selection, of distinction. In a word, a bio-ethics, because it is an "ethics without morality," is not an unmoral activity. Rather, the primary difference between bioethics and bio-ethics is their approach to the question of ethics. Bioethics, following Kantian ethical philosophy, asks how a set of codes or laws can be prescriptively established, such that "evil" can be prevented. A bio-ethics, by contrast, begins by asking, "where does ethics occur"? From this more bottom-up perspective, the question of where ethics occurs also implies a shift in perspective with relation to what counts as a "body," what counts as the "bio" of bioethics.

On a practical note, there is no reason that bioethics and bio-ethics cannot form a counterpoint, an integrated unit dedicated to studying the same topic on two qualitatively different registers. The work in applied bioethics has and is confronting highly complex and extremely difficult issues, ranging from human subjects research in genetic therapy clinical trials to the consideration of the effectiveness of governmental bans on research using human embryonic stem cells harvested from IVF clinics. This work should by all means continue, but in addition to it, we can also imagine a parallel thread in which questions that are critical, philosophical, and in the end pragmatic come to bear on bioethics. This bio-ethical thread would be one in which we seek to critically assess the ontological assumptions and methodological foundations that inform current bioethics approaches. While applied bioethics is primarily concerned with how specific biomedical and biotechnological issues will play themselves out at the level of institutions, funding bodies, and policy decisions, a parallel bio-ethics is primarily concerned with evaluating ethics itself, and exploring possibly transformative, alternative models for furthering and diversifying the current bioethical paradigms.

The bioethics/bio-ethics distinction is not a distinction between theory and practice. Nor is it a split between specific and general problem solving. The distinction between bioethics and bio-ethics is an index of what "ethics" can possibly mean in relation to issues in biomedicine and biotechnology. Ethics need not only signify the practical application of universal moral principles ("good versus evil"); ethics can also be a form of critique, a means of elucidating the limits of what we accept as part of the human condition.

Metadesign

How does this distinction between bioethics and bio-ethics relate to biotechnology as a technology? We have seen how a Deleuzian–Spinozist bio-ethics would foreground the human and nonhuman in its definition of "bodies," something that would seem very appropriate to the biotechnologies of biochips, databases, and lab-grown tissues. But this is still a consideration of ethics at the "output" of biotechnology; what about the "input," those modes of technical development and design that lead to biochips, databases, and in vitro tissue cultures?

In an essay titled "Metadesign," Humberto Maturana raises the question of ethics in relation to the design of living systems.[31] For Maturana, the question is not whether we can or cannot accomplish the design of living systems technically (for putting the question this way makes it a purely technical matter); the question is what role design has in relation to the meanings granted to human activity and human "bodyhood." Maturana takes the common ethical argument of capability versus necessity ("could" versus "should"), and roots the question in the dynamic embodiment of human sub-jects—as both subjects and objects of design processes.

Maturana begins by stating that living systems are "structurally determined systems," where stimuli (internal or external) do not change the system, but rather trigger struc-tural perturbations (perturbations that dynamically define the system as a system). In this sense, living systems are "molecular autopoietic machines," whose primary func-tion is the production of the components that produce the components that form the network that sustains them. In the case of living systems, those components are molec-ular aggregates, and although the system is functionally open (in the exchange of mat-ter and energy), it is autopoietically closed (it has as its product the production of the relations that define it).

As a way of further distinguishing this relation in living systems, Maturana com-pares robots and human beings. Robots are examples of allopoietic systems, in that their goal is something other than themselves, and in that they are constructed not through their own interactions but by human beings. Robots are "ahistorical" inten-tional design. By contrast, human beings have as their defining function (according to Maturana) the production of themselves through time by reason of their organiza-tion. Humans are historical processes of natural development (phylogenetic evolution). The difference between humans and robots thus lies not in their materials (metal versus flesh) but in the way in which their dynamic qualities (change) are character-ized; humans have history, whereas robots are ahistorical.[32]

For Maturana, one (but not the only) result of this ahistoricity that is part of tech-nology is that it can have recursive consequences, when the product of the technology is the human being.[33] Biotechnology is an example, where the ahistorical character of technology is brought to bear on the historical character of the human bodyhood. In biotechnology, the human enframing of technology is employed to enable human be-ings to design themselves as technology.

Whether the ahistorical recursivity of the technological domain or the instance of "biomedia," this process as a recursive process raises the issue of ethics vis-à-vis tech-nological design. As Maturana states:

> the expansion of biotechnology has resulted in an expansion of the knowledge of living
> systems as structurally-determined systems and vice-versa. However it has not expanded
> our understanding of living systems as systems, nor has it expanded our understanding
> of ourselves as human beings.[34]

For Maturana, the issue is not technologization; anything can be designed, once the structure of a system is understood. The issue is integrating ethics with design, or, in this case, integrating bio-ethics with metadesign. Maturana only implies a definition for the term *metadesign*, but he seems to suggest that the question of ethics ("what do we want for ourselves?") be rendered indissociable from the question of design ("what can we make?"). If, generally speaking, "bioethics" involves a consideration of ethical questions pertaining to biomedicine and biotechnology, then "metadesign" involves a reflexive thinking about design as a bioethical endeavor (or as inseparable from bio-ethical concerns).

For Maturana, the particular bodyhood of human beings is essential to any ethical consideration, just as embodiment is essential to human history. This need for an ethics that is directly connected to design or technical questions arises, for Maturana, from our current technoscientific situation, in which the new technologies simply replay the same activities, the same habits, the same ideologies:

> The bodyhood . . . is the condition of possibility of the living system, but the manner of its constitution and continuous realization is itself continuously modulated by the flow of the living system in the domain in which it operates as a totality . . . Therefore, body-hood and manner of operating as a totality are intrinsically dynamically interlaced.[35]

Following Maturana's autopoietic approach, "living systems" are dynamically structured, and metadesign involves the materialization of open-ended systems. The dynamic structuration of those systems enables them to exist in time, to be modulated according to different contexts and situations. They form interactions with other systems, different surrounds, and events and phenomena "out there." Living systems are not simply affected or altered by those agencies "out there." Rather, the particular interactions living systems engage prompt flexibilities in the structures of the living systems, causing the structure to express itself in variance. In other words, the result of the "out there" happening to you is something that was always implicit in your structure, though not manifest. Interactions are not causes and effects, acts and consequences, but rather dynamic restructurations without end.

Thinking about design is one thing; making prescriptive statements across disciplines is quite another. Maturana's point in bringing up the concept of metadesign is a broader, techno-epistemological one: that "we" living systems need to consider human ethics in relation to technology and design activities. But can the point also be made that thinking about design would also require design to think outside of traditional human-tool categories? A more radical question would be, can design in relation to living systems think outside of anthropomorphism?

From one perspective, metadesign can also become a program for a new outlook on "bio-design," or the relation between design and living systems. A-life and bottom-up approaches in AI have already been exploring such issues, but rarely do the same concerns with complexity cross over into biotechnology and bioinformatics. Thus meta-design as a prescriptive statement, as a program, would attempt to intersect these fields—already related in their use of biological tropes ("genetic algorithms," etc.)—and in doing so would raise the question of "the human" in relation to this particular type of design at the molecular-informatic level.

But bio-ethics should not simply be limited to the molar level of designing for the human subject, for the subject is much more, or in excess of, the liberal-humanist notion of an "individual." Although this individual is configured as a "free agent" in the democratic marketplace of health insurance, we might ask how subjects are articulated, not only through bioethics discourse, but also through institutions such as health care, hospitals, clinics, and technical instances, such as diagnostics, product development, and advanced computation. In addition to the traditional philosophical ethics and applied ethics at work, we also need modes of diversifying the ethical paradigms currently in place. Again, this is not to suggest that a new paradigm would overthrow an earlier one, but that ethics, as a concern with the other, would seem to require "other" modes of thinking about ethics, modes that may very well take us out of the familiar terms of "individual," "society," or "moral law."

Given our discussion thus far, there appear to be two primary challenges for a bioethics (a nonhuman ethics based on "bodies" as capacities for affect) that would operate through metadesign (ethical practice as design). The first challenge is, as already stated, thinking outside of the liberal-humanist standpoint. This is in part a theoretical difficulty in "thinking otherwise," but it is also rooted in a range of disciplines, knowledges, and institutions. The second challenge, connected to the first, is that that model of the subject we take for granted is not merely a matter of habit, but is something embedded in our very legal system. Patients who have individual rights, physicians who are individually accountable, research teams and companies who own patents on certain techniques or inventions—all these modes of bioethics are inscribed in our legal system, which facilitates the modes of individuation particular to juridical subjects.

These two challenges can be combined into a single comment: we can think of a nonhuman bio-ethics as much as we want, but it will never become a law. This single comment has a twofold meaning. Of course it will never become a law, for our modern legal system excludes modes of individuation that are not particular to human subjects as autonomous agents. In fact, one way to gauge the level of departure of alternative bioethical models is to ask if and how they might be incorporated into the legal system. So then, bioethics, as a discipline and institution, implicitly excludes bioethics, as a mode of critical inquiry. The real problematic, then, is first pointing to the precise relations of incommensurability between these two perspectives (bioethics and bio-ethics), and then asking how bioethics can be made more flexible, to accommodate the strange hybrids that emerging biotechnologies are generating. The approach of bio-ethics might be the following question: what is "the other" of ethics?[36]

Notes

1. What Is Biomedia?

The heading "New ways, new ways, I dream of wires" is from the song "I Dream of Wires," by Gary Numan, from the album *Telekon* (Beggar's Banquet, 1980).

1. This is just one of many examples of the often cagey relationships between academia and industry that have been one of the defining aspects of biotechnology as a discipline. While academic and government-funded researchers were limited to highly specific contexts (e.g., discarded fetal tissue from in vitro fertilization [IVF] clinics), biotech companies specializing in stem cell research and "regenerative medicine" (most notably Geron) promised that stem cells would not only pave the way to curing neurodegenerative disorders such as Parkinson's disease, but would also make possible "off-the-shelf organs." For more on regenerative medicine see the *Scientific American* special issue "Growing New Organs" (April 1999).

2. The U.S. government's National Bioethics Advisory Commission's statements on the issue are available on its Web site, at http://www.bioethics.gov.

3. Briefly, genomics is the quantitative analysis of entire genomes. Some branches of genomics specialize in a single species, such as the yeast, fruit fly, mouse, or human genomes. Other approaches compare similar genomes (say, mouse and human) in order to seek common bases for disease (hence the common use of mice in cancer genetics research). Proteomics is the study of the proteins produced by a genome. Many suggest that proteomics is impossible without genomics, implying that genomics provides the foundation on which proteomics works. Proteomics has the immense challenge of articulating how the genome specifies the production of specific proteins, and what those proteins do. There are several initiatives under way for a "human proteome project," although, given the number of possible proteins in the human body (there is no definite count, but it is exponentially greater than the number of genes), this is a daunting endeavor. Both genomics and proteomics intersect with industry in pharmacogenomics, which is the application of genomic and proteomic knowledge toward drug design. This is the primary reason for "big pharma" or large pharmaceutical company interest in genomics and proteomics, as possible techniques for "targeting" key disease genes.

4. Ken Howard, "The Bioinformatics Gold Rush," *Scientific American* (July 2000): 58–63, and Aris Persidis, "Bioinformatics," *Nature Biotechnology* 17 (1999): 828–30.

5. The basic bioinformatics techniques of homology modeling are described in some detail in Andreas Baxevanis and B. F. Francis Ouellette, eds., *Bioinformatics: A Practical Guide to the Analysis*

of Genes and Proteins (New York: Wiley-Liss, 2001). The chapters dealing specifically with protein prediction also contain links to Web-based tools commonly used in techniques such as homology modeling, secondary structure prediction, and molecular modeling.

6. Chapter 4 on DNA computing goes into further detail on the varieties of biocomputing approaches. However, for descriptions of DNA computing, see Alan Dove, "From Bits to Bases," *Nature Biotechnology* 16 (1998): 830–32; and Antonio Regalado, "DNA Computing," *MIT Technology Review* (May/June 2000): http://www.techreview.com/articles/mayoo/regalado.htm.

7. Alan Turing's famous paper describing his hypothetical "universal machine" is often cited as a key text in the history of computer science, for, at an early stage in the development of mainframe digital computers, it articulated the limits of computing. The universal machine was an imaginary construct, a computer that could be fed the instructions of another machine, and that would then run just like that machine (analogous to emulators in modern PCs). However, when this universal machine was fed its own instructions, it would be unable to get outside of itself in order to compute, caught in a positive feedback loop (unable to emulate itself because it already is itself). See Alan Turing, "On Computable Numbers with an Application to the *Entscheidungsproblem*," *Proceedings of the London Mathematical Society* 2.42 (1936): 230–67.

8. The theme of the technologically enabled "brain in a vat" is one that runs through much science fiction. It combines the privileging of "mind" (often conflated in science fiction with the brain) and intelligence over the limitations and contingencies of the "flesh," or what cyberpunk science-fiction authors often called the "meat." Its most recent manifestation is in cyberspace, what William Gibson referred to as the "nonspace of the mind." See William Gibson, *Neuromancer* (New York: Ace, 1984).

9. Walter Benjamin, "The Work of Art in the Age of Mechanical Reproduction," in *Illuminations,* ed. Hannah Arendt, trans. Harry Zohn (New York: Schocken, 1968), 217–51.

10. Martin Heidegger, "The Question concerning Technology," in *The Question concerning Technology and Other Essays* (New York: Harper, 1977), 1–35.

11. Marshall McLuhan, *Understanding Media* (Cambridge: MIT Press, 1995).

12. The most cogent analysis of posthumanist and/or Extropian thinking is that of Katherine Hayles, who shows how posthumanist futurists such as Hans Moravec and Ray Kurzweil revisit transcendental narratives through the high-tech, informatic tropes of "pattern" instead of "presence." Similarly, the writings of Arthur and Marilouise Kroker have focused, through a strategic "panic theory," on the ways in which technological intensification affects shifting cultural attitudes toward the body. Finally, Haraway's well-known essay on the cyborg has played a role in helping to facilitate thinking about technology and cultural difference rooted in the body (specifically in relation to race, gender, and the economics of the "integrated circuit"). See Katherine Hayles, *How We Became Posthuman* (Chicago: University of Chicago Press, 1999); Arthur and Marilouise Kroker, eds., *Digital Delirium* (New York: St. Martin's Press, 1997); and Donna Haraway, *Simians, Cyborgs, and Women* (New York: Routledge, 1991).

13. Jay Bolter and Richard Grusin, *Remediation: Understanding New Media* (Cambridge: MIT Press, 1999), 98.

14. Ibid., 5.

15. The first four of the characteristics are immanently technical with broader cultural impact. "Numerical representation" begins as a descriptive notion (that new media objects exist in digital format as numbers) and is extended into a proscriptive notion (that anything which can be digitally encoded is amenable to the manipulations of new media). "Modularity" describes the ability of new media elements to be combined and recombined in a variety of ways: QuickTime movies in Flash files in Netscape browsers on Mac OSX platforms. "Automation," as its name implies, points to the continued interest in software development and new media generally to "black box" much of what happens behind the scenes. A series of laborious steps in rendering a 3D object can be automated using a single menu option in a graphical interface. "Variability" is a characteristic that describes the metamorphic nature of new media objects. Web pages are not static, but can be updated, redesigned, relocated, downloaded, and viewed in a variety of ways. Although these characteristics are implicit in

the discussion of biomedia here, the primary focus is on Manovich's fifth characteristic of "transcoding." For more on these characteristics, see Lev Manovich, *The Language of New Media* (Cambridge: MIT Press, 2002), 27–48.

16. Ibid., 46.

17. Bolter and Grusin, *Remediation*, 238.

18. The term "techniques of the body" refers to an essay of the same name by Marcel Mauss. Mauss's anthropological perspective analyzes the ways in which everyday actions such as walking, running, sleeping, swimming, playing, and climbing are not abstract but contextualized in social settings, where the biological, the psychological, and the social intersect. Modes of walking, for instance, are contextualized one way in dense urban settings, and another in country settings. Mauss states, "I call 'technique' an action that is effective and traditional." The role of training, education, and "tradition" (normal versus abnormal) are key elements in Mauss's view of how the body becomes a set of learned "techniques." See Marcel Mauss, "Techniques of the Body," in *Zone 6: Incorporations*, ed. Jonathan Crary and Sanford Kwinter (New York: Zone Books, 1992), 454–77.

19. We can distinguish this further by those uses of bioscience for medical ends, such as regenerative medicine (the use of the body's own cellular regenerative capacities to develop novel therapies); those uses for diagnostic ends, such as biochips (DNA on chips used for genetic analysis); those uses for computational ends, such as DNA computing (described earlier); and those uses for engineering, such as biomimcry, of the application of biological principles toward industrial and mechanical engineering.

20. Robert Bud, *The Uses of Life: A History of Biotechnology* (Cambridge: Cambridge University Press, 1993).

21. This subject anxiety is often demonstrated in science-fiction film, such as *The Sixth Day*, *X-Men*, and *Gattaca*, all of which position heroic, liberal humanist subjects against the hegemony of biotech corporatism or governmentalism.

22. Jean Baudrillard, *Seduction*, trans. Brian Singer (New York: St. Mark's, 1990).

23. Donna Haraway, *Modest.Witness@Second_Millennium* (New York: Routledge, 1997), 131–67.

24. Evelyn Fox Keller, *Refiguring Life: Metaphors of Twentieth-Century Biology* (New York: Columbia University Press, 1995); Catherine Waldby, *The Visible Human Project: Informatic Bodies and Posthuman Medicine* (New York: Routledge, 2000).

25. Judith Butler, *Bodies That Matter* (New York: Routledge, 1993), 90. In spite of the many books that have recently been published on cultural theory and "the body," Butler's text remains one of the most thorough theoretical interrogations of the relationship between discourse and materiality. Butler consistently produces a tension-filled zone in attempting to grapple with the materiality/discourse relation. As an instance of signification, the body is articulated through language (is only intuited/understood through conceptual paradigms) but is never fully of language (it never totally fullfills its referent, or it produces a surplus as part of its meaning-making process that retains a position exterior to the language in which it is articulated). With the body, the function of language, as that which marks differentiations of signifiers and signifieds in order to articulate more clearly its object, is here undermined by the signification of an incomplete referent whose posited ontology is that it exceeds language. But the very materiality of language (both as sensory/phenomenal signs and as its material-political effects) indicates that (1) materiality is never purely outside of language and (2) this contingency does not negate the real material effects of materiality not being distinct from language. Every attempt to "get at" materiality must go through language, whose operation is fully material. Thus language and materiality are not opposed or incommensurably different, "for language both is and refers to that which is material, and what is material never fully escapes from the process by which it is signified" (ibid.).

26. Maurice Merleau-Ponty, *The Phenomenology of Perception* (New York: Routledge, 1992).

27. Drew Leder, *The Absent Body* (Chicago: University of Chicago Press, 1990).

28. "Corporealization" is a term Donna Haraway uses to discuss how metaphors get mistaken for "non-tropic things in themselves" (Haraway, *Modest.Witness@Second_Millennium*, 141–48).

29. In computer programming terms, "compiling" is the process in which a program is translated from human-readable form (e.g., a higher-level programming language such as Java or C++) into a "machine-readable" form of ones and zeros.

30. Although this point is arguable, Heidegger's overall approach of eliciting the "essence" of technology as an activity begins from the premise of the ontological differentiation between human subject/body and technical object/activity. Whether it is in the premodern mode of "revealing" or "bringing forth," or in the modern, instrumentalized mode of "setting upon," Heidegger's questioning of technology centers on the point of contact between hand and tool, whose essence is a particular mode of activity in relation to the material ("natural") world. See Heidegger, "The Question concerning Technology," 4–12.

31. The best examples of the latter two technological tropes—that of the prosthetic and that of displacement—is not theory but science fiction. Cyborg narratives, such as Fredrick Pohl's *Man Plus,* explore how prosthetic technologies not only augment the human subject, but in the process also transform human subjectivity. Science-fiction films such as *Robocop* replay this theme. Similarly, intelligent machine narratives such as Brian Aldiss's *Supertoys Last All Summer Long* pose the familiar question of the point at which machines begin to compete with the human, a theme played out famously in the film *2001: A Space Odyssey.* See Fredrick Pohl, *Man Plus* (London: Gollancz, 1972); and Brian Aldiss, *Supertoys Last All Summer Long* (New York: Griffin, 2001).

32. Again, Heidegger's notion of "enframing" is useful here, in the sense that it highlights the way in which "technology" is a particular mode of relating to the world. However in his distinction between premodern crafts and modern industrialism, Heidegger tends to evoke a kind of romanticism for one relating (based on a more anthropomorphic relation, mediated by the "hand" of the craftsperson) over another (based on an empty application of rational principles to the resourcing of the natural world). A critical reading working from Heidegger's text would have to, in addition, "question" the role of nature and the biological as well as the artificial means for relating, and in doing so, address the tendency to equate mediation with alienation. See Heidegger, "The Question concerning Technology," 19–23.

33. PCR (polymerase chain reaction) is a standard laboratory technology for rapidly replicating desired sequences of DNA. PCR machines take in a sample and put it through cycles of cooling and heating, which causes DNA to denature into single strands, and then replicate into duplicate strands. Gel electrophoresis—an earlier form of gene sequencing—is another standard lab technique, which involves passing a DNA sample through a gel-like medium using electrical charges. A researcher can sequence the DNA based on the combination of fragment length and its comparison to a known sequence. Recombinant DNA is a basic technique in genetic engineering, which utilizes special "cutting" molecules known as restriction endonucleases to snip a DNA sequence at a desired spot, so that a new sequence can be inserted into the DNA.

34. Base pair complementarity, also known as "Watson–Crick complementarity," is the principle that complementary base pairs in DNA and RNA will, under normal conditions, always bind to each other: adenine (A) always binds with thymine (T), guanine (G) always binds with cytosine (C). This principle, articulated by Watson and Crick in the 1950s, is one of the most common processes utilized in biomedia. See James Watson and Francis Crick, "General Implications of the Structure of Deoxyribonucleic Acid," *Nature* 171 (1953): 964–67.

35. The contextualization of biomedia as a "protocol" of encoding, recoding, and decoding was presented at a Crossroads in Cultural Studies conference (Birmingham University, 21–25 June 2000), as part of a panel on posthumanism. There the specific example referred to was tissue engineering, and the ways in which cellular regeneration was recontextualized as biological optimization. The concept of protocol has also, more recently, greatly benefited from the work of Alex Galloway, who has elaborated how computer networking forms a materialization of "how control exists after decentralization." See Eugene Thacker, "Data Made Flesh: Biotechnology and the Discourse of the Posthuman," *Cultural Critique* 53 (spring 2002); Alex Galloway, "Protocol, or, How Control Exists after Decentralization," *Rethinking Marxism* 13.3/4 (2001).

36. As Hayles states, "information in fact derives its efficacy from the material infrastructures it appears to obscure" (*How We Became Posthuman*, 28).

37. The most well known of such formulations is in Aristotle's notion of "entelechy," where formless matter exists in a teleological framework so as to be realized in form. See Aristotle, *De Anima* (New York: Penguin, 1986), book 2, chapters 1–3.

38. See Claude Shannon and Warren Weaver, *The Mathematical Theory of Information* (Chicago: University of Illinois Press, 1965).

39. Such a question, which at first seems abstract, is implicit in current debates over intellectual property, the patenting of biological materials, the privatization of genome databases, and the state of open-source movements in the biotech software industry.

40. It is significant to note that, in molecular biology, the terms *transcription* and *translation* have been used for some time to describe the process by which DNA is "read" onto RNA, and the process by which RNA "writes" an amino acid chain that becomes a protein. The informatic and textual tropes of molecular biology exist on a level much more specific and detailed than the broad notion of a genetic "code." However, as several cultural critics have noted, the application of such tropes often problematizes rather than clarifies the situation. See, for example, Keller, *Refiguring Life*, 79–118.

41. Open source generally refers to any instance in which the source code for a software program, operating system, or other type of computer-based media object is made available to the public. Most often communities of developers organize around a particular project, in a collaborative environment. The open-source movement has often been lauded as the latest instance of utopianism on the Internet, with the Linux operating system being the most famous example. However, although software code may be free, it has also spawned an industry surrounding it, including technical support services, book publishing, and cloned for-sale applications. See Adrian MacKenzie, "Open Source Software: When Is It a Tool? What Is a Commodity?" *Science as Culture* 10.4 (2001): 541–52.

42. For more on this technique, see Cynthia Gibas and Per Gambeck, *Developing Bioinformatics Computer Skills* (Cambridge: O'Reilly, 2000).

43. Both tools are freely accessible through the National Center for Biotechnology Information's Web site, at http://www.ncbi.nlm.nih.gov.

44. The ExPASy (Expert Protein Analysis System) set of tools is freely available through the Swiss Bioinformatics Institute, at http://www.expasy.ch.

45. Although Jean Baudrillard suggests that the very notion of a genetic "code" shows how informatics is an exemplary case of simulation, it is also manifested in techniques such as human cloning, which Baudrillard refers to as an ontological "forgetting of death." What Baudrillard overlooks, however, is the way in which fields like bioinformatics constitute modes of technically impelling biological life to be even more "biological." See Jean Baudrillard, *The Vital Illusion*, trans. Julia Witwer (New York: Columbia University Press, 2000).

46. Regenerative medicine is a broad name for research fields including tissue engineering, gene therapy techniques, and stem cell research. The term has been promoted by William Haseltine, CEO of Human Genome Sciences, as the application of biotechnology to enable more long-term, robust therapies of genetically based disorders that affect cell and tissue function. In popular terms, this hints at improved immune systems and "lab-grown organs," though the feasibility of these has yet to be shown.

47. The connection between the genetic code and cryptography has been made several times. Among the earliest instances was the initial "cracking of the genetic code" by Heinrich Matthai and Marshall Nirenberg in the early 1960s. See Lily Kay, *Who Wrote the Book of Life? A History of the Genetic Code* (Standford, Calif.: Stanford University Press, 2000), 246–56.

48. For more on pharmacogenomics and rational drug design, see W. E. Evans and M. V. Relling, "Pharmacogenomics: Translating Functional Genomics into Rational Therapeutics," *Science* 286 (15 October 1999): 487–91; Jennifer van Brunt, "Pharma's New Vision," *Signals Magazine* (1 June 2000): http://www.signalsmag.com; Geoffrey Cowley and Anne Underwood, "A Revolution in Medicine," *Time* (10 April 2000): 58–67.

49. This "silver bullet" approach was pioneered by French Anderson, who treated a little girl with a rare single gene mutation disorder with this type of gene therapy. See French Anderson, "The ADA Human Gene Therapy Clinical Protocol," *Human Gene Therapy* 1 (1990): 327–62.

50. One example is the 1999 death of a teenage boy undergoing a gene therapy clinical trial, because of side effects resulting from the novel genes on other, nonrelated, organ systems in his body. The death not only raised issues of human subject research in gene therapy, but also marked, perhaps permanently, the cultural connotations of gene therapy.

51. The question "What can a body do?" is posed by Gilles Deleuze in his reading of the work of Spinoza. Deleuze points to Spinoza's ontology of bodies by highlighting its particularly operational character. If Spinoza begins by aligning "nature" and "God" (one substance for all attributes), then the concept of matter that follows from this implies a "parallelism" between body and mind (in contradistinction to Descartes's dualism). The result of this—highlighted by Deleuze—is a twofold conception of "bodies" as constituted by relations of motion and rest, and as constituted by affects. This latter notion becomes especially relevant in Spinoza's discussion of ethics. The force of this Deleuzian–Spinozist question in relation to biomedia is to displace the question of "being" with that of "becoming," inasmuch as the latter presumes a situation in which a body cannot be defined without considering its context and operation in time as a dynamic entity. See Gilles Deleuze, *Spinoza: Practical Philosophy* (San Francisco: City Lights, 1988).

52. See Hayles, *How We Became Posthuman*. Also, for uncritical, pro-posthumanist accounts, see Ray Kurzweil, *The Age of Spiritual Machines* (New York: Penguin, 1999); Hans Moravec, *Robot: Mere Machine to Transcendent Mind* (Oxford: Cambridge University Press, 1999); and Max More, "The Extropian Principles: A Transhumanist Declaration," *Extropy.org* (1999): http://www.extropy.org.

53. There is a specific discourse within molecular biology that extrapolates from the specifics of the genetic code to its general meanings for teleology, determinism, and reductionism. This can be seen through popular books by scientists that either have the title "What Is Life?" or whose title is some comment on this question. See, for instance, Erwin Schrödinger, *What Is Life?* (Cambridge: Cambridge University Press, 1967); Francis Crick, *Life Itself: Its Origin and Nature* (New York: Simon and Schuster, 1981); and François Jacob, *The Logic of Life* (New York: Pantheon, 1974).

54. One important example is Critical Art Ensemble, which, for a number of years, has combined bioactivism in practice with publications, talks, and performances that address the relations between the body, biotech, and global capitalism. See Critical Art Ensemble, *Flesh Machine* (Brooklyn: Autonomedia, 1998), and *Molecular Revolution* (Brooklyn: Autonomedia, 2002). Also see Haraway, *Modest.Witness@Second_Millennium*.

55. See Dorothy Nelkin and Susan Lindee, *The DNA Mystique: The Gene as a Cultural Icon* (New York: W. H. Freeman, 1995); and Jon Turney, *Frankenstein's Footsteps: Science, Genetics, and Popular Culture* (New Haven: Yale University Press, 1998).

56. The term *fiction science* is used by Pierre Baldi, a bioinformatician, to describe the link between actual science research and the ways in which projected application gets taken up both in the scientific community and in the popular media. See Pierre Baldi, *The Shattered Self* (Cambridge: MIT Press, 2000).

2. Bioinformatics

The heading "We're in the building where they make us grow" is from the song "Metal," by Gary Numan, from the album *The Pleasure Principle* (Beggar's Banquet, 1979).

1. Greg Egan, *Diaspora* (New York: HarperCollins, 1998).
2. Ibid., 6.
3. Ibid., 12. Egan's imagining of this process consciously mixes the gendered tropes of biological reproduction with informatic processes of differential replication of software code, akin to the use of cycling genetic algorithms, where particular patterns of code attempt to "adapt" to a given computational problem.

4. Aside from the informatically based sentient citizens, there are also "gleisners," or citizens who prefer to inhabit robotic, hardware bodies, rather than existing in an abstract information space. The gleisners replay Cartesian dualism, but through the filter of AI and robotics. There are also "fleshers," or biologically based human beings who have fully incorporated genetic engineering. Fleshers who refuse any form of genetic modification are known as "statics," and fleshers who adopt more extreme modes of modification (the Dream Apes and aquatic humanoids) are referred to as "exuberants." This is only the beginning, however, of a complex novel that eventually comes across computational alien kelp-like creatures who may or may not be sentient, and n-dimensional "hermits," alien mollusk-like creatures who, in the ultimate act of Darwinian adaptation, occupy several dimensions at once, and devour the different material worlds around them.

5. For more on bioinformatics, see Ken Howard, "The Bioinformatics Gold Rush," *Scientific American* (July 2000): 58–63; Nathan Goodman, "Biological Data Becomes Computer Literate: New Advances in Bioinformatics," *Current Opinion in Biotechnology* 31 (2002): 68–71; Aris Persidis, "Bioinformatics," *Nature Biotechnology* 17 (August 1999): 828–30; Bernhard Palsson, "The Challenges of in silico Biology," *Nature Biotechnology* 18 (November 2000): 1147–50; and Arielle Emmett, "The State of Bioinformatics," *Scientist* 14.3 (27 November 2000): 1, 10–12, 19.

6. This proto-informatic approach to the biological/medical body can be seen in Michel Foucault's work, as well as in analyses inspired by him. Foucault's analysis of the medical "gaze" and nosology's use of elaborate taxonomic and tabulating systems are seen to emerge alongside the modern clinic. Such technologies are, in the later Foucault, constitutive of the biopolitical view, in which power no longer rules over death, but rather impels life, through the institutional practices of the hospital, health insurance, demographics, and medical practice generally. See Michel Foucault, *The Birth of the Clinic: An Archaeology of Medical Perception* (New York: Vintage, 1973); Michel Foucault, "The Birth of Biopolitics," in *Ethics: Subjectivity and Truth*, ed. Paul Rabinow (New York: New Press, 1994): 73–81. For collections of recent analyses that take up Foucault's approach, see Colin Jones and Roy Porter, eds., *Reassessing Foucault: Power, Medicine and the Body* (New York: Routledge, 1998); and Alan Petersen and Robin Bunton, eds., *Foucault, Health, and Medicine* (New York: Routledge, 1997).

7. Fred Sanger, "The Structure of Insulin," in *Currents in Biochemical Research*, ed. D. E. Green (New York: Wiley Interscience, 1956).

8. R. W. Holley et al., "The Base Sequence of Yeast Alanine Transfer RNA," *Science* 147 (1965): 1462–65.

9. A number of researchers began to study the genes and proteins of model organisms (often bacteria, roundworms, or the Drosophila fruit fly), not so much from the perspective of biochemical action, but from the point of view of information storage. For an important example of an early database-approach, see Margaret Dayhoff, *Atlas of Protein Sequences and Structure* (Silver Springs, Md.: National Biomedical Research Foundation, 1966).

10. This, of course, was made possible by the corresponding advancements in computer technology, most notably in the PC market. See E. E. Abola et al., "Protein Data Bank," in *Crystallographic Databases*, ed. F. H. Allen et al. (Cambridge: Data Commission of the International Union of Crystallography, 1987), 107–32; and A. Bairoch and R. Apweile, "The SWISS-PROT Protein Sequence Data Bank," *Nucleic Acids Research* 19 (1991): 2247–49.

11. Initially, the HGP was funded by the U.S. Department of Energy and the National Institutes of Health, and had already begun forming alliances with European research institutes in the late 1980s (which would result in HUGO, or the Human Genome Organization). As sequencing endeavors began to become increasingly distributed to selected research centers and/or universities in the United States and Europe, the Department of Energy's involvement lessened, and the NIH formed a broader alliance, the International Human Genome Sequencing Effort, with the main players being MIT's Whitehead Institute, the Welcome Trust (U.K.), Stanford Human Genome Center, and the Joint Genome Institute, among many others. This broad alliance was challenged when the biotech company Celera (then the Institute for Genome Research) proposed its own genome project funded by the corporate sector. See Daniel Kevles and Leroy Hood, eds., *The Code of Codes: Scientific and So-*

cial Issues in the Human Genome Project (Cambridge: Harvard University Press, 1992); and, for a popular account, see Matt Ridley, *Genome* (New York: Perennial, 1999).

12. From the Bioinformatics.org Web site: http://www.bioinformatics.org.

13. Silico Research Limited, "Bioinformatics Platforms," Research Report/Executive Summary (November 2000): http://www.silico-research.com.

14. Oscar Gruss/Jason Reed, "Trends in Commercial Bioinformatics," *Biotechnology Review* (13 March 2000): http://www.oscargruss.com.

15. Press releases and news stories are accessible through the company Web sites. See also Declan Butler, "Computing 2010: From Black Holes to Biology," *Nature* 402 (2 December 1999): C67–70.

16. For more on the technique of pairwise sequence alignment, see Cynthia Gibas and Per Gambeck, *Developing Bioinformatics Computers Skills* (Cambridge: O'Reilly, 2000).

17. For an overview of various bioinformatics techniques, see Andreas Baxevanis and B. F. Francis Ouellette, eds., *Bioinformatics: A Practical Guide to the Analysis of Genes and Proteins* (New York: Wiley-Liss, 2001).

18. These and other terms are defined on-line by the National Human Genome Research Institute (NHGRI), in the NHGRI's glossary of molecular biology, at http://www.nhgri.nih.gov/DIR/VIP/Glossary/pub_glossary.cgi.

19. Erwin Schrödinger, *What Is Life?* (Cambridge: Cambridge University Press, 1967).

20. Francis Crick, *Life Itself: Its Origin and Nature* (New York: Simon and Schuster, 1981). See also Francis Crick, "The Genetic Code," in *Nobel Lectures in Molecular Biology, 1933–1975,* ed. David Baltimore (New York: Elsevier, 1977), 205–13; and Francis Crick, "The Genetic Code—Yesterday, Today, and Tomorrow," *Cold Spring Harbor Symposia on Quantitative Biology* 31 (1966): 3–9.

21. George Beadle and Muriel Beadle, *The Language of Life* (New York: Doubleday, 1966).

22. François Jacob, *The Logic of Life: A History of Heredity* (New York: Pantheon, 1974); Jacques Monod, *Chance and Necessity: An Essay on the Natural Philosophy of Modern Biology* (London: Fontant/Collins, 1974).

23. Lily Kay, *Who Wrote the Book of Life? A History of the Genetic Code* (Stanford, Calif.: Stanford University Press, 2000). See also Lily Kay, "Cynernetics, Information, Life: The Emergence of Scriptural Representations of Heredity," *Configurations* 5.1 (1997): 23–91.

24. Susan Aldridge, *The Thread of Life: The Story of Genes and Genetic Engineering* (Cambridge: Cambridge University Press, 1998).

25. For key articles on recombinant DNA, see S. N. Cohen, A. C. Chang, H. W. Bayer, and R. B. Helling, "Construction of Biologically Functional Bacterial Plasmids *in vitro*," *Proceedings of the National Academy of Sciences* 70 (1973): 3240–44; and A. C. Y. Chang and S. N. Cohen "Genome Construction between Bacterial Species *in vitro*: Replication and Expression of Staphylococcus Plasmid Genes in Escherichia coli," *Proceedings of the National Academy of Sciences* 71 (1974): 1030–34. The work of Cohen and Boyer's teams resulted in a U.S. patent, "Process for Producing Biologically Functional Molecular Chimeras" (#4237224), as well as the launching of one of the first biotech start-ups, Genentech, in 1980. See Cynthia Robbins-Roth, *From Alchemy to IPO: The Business of Biotechnology* (New York: Perseus, 2000).

26. Documents on the history and development of the U.S. Human Genome Project can be accessed through its Web site: http://www.ornl.gov/hgmis.

27. It would be inaccurate to say that metaphor simply disappears, and the "real thing" emerges; it is the way in which metaphor is positioned vis-à-vis biological materiality that has changed. Cultural theorists such as Donna Haraway note how metaphor is never separate from materiality; her analyses of "genetic fetishism" and the cultural syntax and semantics of biotechnology aim to demonstrate how metaphor is mobilized toward self-transparency. The difference being articulated here is not that metaphor does not "matter," but that it provides the occasion for a mode of technical implementation in the biologization of technology. On the informatic tropes of molecular biology, see Evelyn Fox Keller, *Refiguring Life: Metaphors of Twentieth-Century Biology* (New York: Columbia University Press, 1995); Donna Haraway, *Modest.Witness@Second_Millennium* (New York: Routledge, 1997); Robert Pollack, *Signs of Life: The Language and Meaning of DNA* (Boston: Houghton Mifflin,

1994); Soraya de Chadarevian, "Sequences, Conformation, Information: Biochemists and Molecular Biologists in the 1950s," *Journal of the History of Biology* 29 (1996): 361–86; Thomas Fogle, "Information Metaphors and the Human Genome Project," *Perspectives in Biology and Medicine* 38.4 (1995): 535–47; and Marcel Weber, "Representing Genes: Classical Mapping Techniques and the Growth of Genetical Knowledge," *Studies in the History and Philosophy of Biological and Biomedical Sciences* 29.2 (1998): 295–315.

28. Alan Dove, "From Bits to Bases: Computing with DNA," *Nature Biotechnology* 16 (September 1998): 830–32.

29. For summaries of these perspectives, see Fritjof Capra, *The Web of Life: A New Understanding of Living Systems* (New York: Doubleday, 1997); and John Brockman, ed., *The Third Culture* (New York: Touchstone, 1995).

30. Pierre Baldi, *The Shattered Self* (Cambridge: MIT Press, 2000).

31. Ibid., 3.

32. See C. E. MacKenzie, *Coded Character Sets: History and Development* (Reading, Mass.: Addison-Wesley, 1980).

33. During the latter part of the twentieth century, this position was simultaneously reflected in certain strands of cultural theory, computer science research, and science fiction. See Jean Baudrillard, *Simulations* (New York: Semiotext[e], 1983); Hans Moravec, *Mind Children: The Future of Robot and Human Intelligence* (Oxford: Cambridge University Press, 1988); and William Gibson, *Neuromancer* (New York: Ace, 1984). More recent perspectives have critiqued these views by locating the digital and virtual in cultural contexts, where gender, sexuality, and race play constitutive roles. However, the predominant discursive mode in these critiques has been at the level of representation, and has rarely questioned the viability of the real/digital, or better, biological/digital, boundary itself. See Sadie Plant, *Zeros + Ones: Digital Women and the New Technoculture* (New York: Doubleday, 1997); Allucquère Rosanne Stone, *The War of Desire and Technology at the Close of the Mechanical Age* (Cambridge: MIT Press, 1996); Beth Kolko, Lisa Nakamura, and Gilbert Rodman, eds., *Race in Cyberspace* (New York: Routledge, 2000).

34. At http://www.bioinformatics.org.

35. "Translation" is used here in several senses: first, in a molecular biological sense, in which DNA "transcribes" RNA, and RNA "translates" a protein; second, in a linguistic sense, in which some content is presumably conserved in the translation from one language to another (a process that is partially automated in on-line translation tools); third, in a computer science sense, in which translation is akin to "portability," or the ability for a software application to operate across different platforms (Mac, PC, Unix, SGI).

36. Claude Shannon and Warren Weaver, *The Mathematical Theory of Communication* (Chicago: University of Illinois Press, 1965).

37. Despite the high turnover rate of bioinformatics software tools, such changes are underwritten by a push toward standardization of database formats, GUI-style interfaces, and the marketing of entire "suites" of tools (hardware and software) for performing biological research. See Paolo Saviotti, Marie-Angele de Looze, Sylvie Michell, and David Cathorine, "The Changing Marketplace of Bioinformatics," *Nature Biotechnology* 18 (December 2000): 1247–49; and Rob James, "Differentiating Genomics Companies," *Nature Biotechnology* 18 (February 2000): 153–55.

38. Bruno Latour, *Pandora's Hope: Essays on the Reality of Science Studies* (Cambridge: Harvard University Press, 1999); Bruno Latour, *The Pasteurization of France* (Cambridge: Harvard University Press, 1988).

39. Certainly, the wet lab of molecular biology has not been without its examples of radical change as a result of technical development—the emergence of PCR (polymerase chain reaction—essentially a DNA Xerox machine) is one such example. See Paul Rabinow, *Making PCR* (Chicago: University of Chicago Press, 1996). The point I want to highlight in relation to bioinformatics is the interdisciplinary nature of bioinformatics. A number of bioinformatics companies were first IT companies, and have created spin-off subdivisions for the life sciences (Sun, Motorola, IBM). This would seem to imply that, in bioinformatics, the cycles of the IT industries set the terms for techno-

204 Notes to Chapter 2

logical advance in the life sciences. The increasing ubiquity of bench-top gene-sequencing computers is an indicator of this kind of change.

40. S. F. Altschul, W. Fish, W. Miller, E. W. Myers, and D. J. Lipman, "Basic Local Alignment Search Tool," *Journal of Molecular Biology* 215 (1990): 403–10.

41. GenBank is the main repository of genome data for the federally funded Human Genome Project, and it also holds genomes for various other model organisms, from bacteria, to yeast, to the roundworm, to mice, to humans. GenBank is accessible at: http://www.ncbi.nlm.nih.gov/Genbank/index.html.

42. UNIX is an operating system developed in the 1970s and 1980s at Bell Labs, and subsequently expanded by programmers such as Bill Joy at Sun and University of California, Berkeley. Unix is particularly adept at handling strings of data, and helped to extend the popularity of the "workstation," or total computing environment for specific user-defined tasks. There are many different "flavors" of Unix today, but most involve the use of a "command line" or text-based prompt, into which the user can type abbreviated commands ("cd" for changing the current directory, "rm file_x" for removing file x, etc.). Unix workstations were particularly favored by universities because, by paying a nominal fee, programmers could obtain the Unix source code, and configure a workstation as they pleased. Because of this lineage of university-based workstation research, and because of its ability in manipulating strings, Unix is still used by many bioinformaticians, although newer applications, in an appeal to nonprogramming biologists, are taking the graphical user interface (GUI) approach. See Paul Ceruzzi, *A History of Modern Computing* (Cambridge: MIT Press, 1998), 281–90.

43. To access the Web version of BLAST, go to http://www.ncbi.nlm.nih.gov:80/BLAST. To access the UCSD Biology Workbench, go to http://workbench.sdsc.edu.

44. A key differentiation in bioinformatics as a field is between research on "sequence" and research on "structure." The former focuses on relationships between linear code, whether nucleic acids/DNA or amino acids/proteins. The latter focuses on relationships between sequence and their molecular-physical organization. If a researcher wants to simply identify a test sample of DNA, the sequence approach may be used; if a researcher wants to know how a particular amino acid sequence folds into a three-dimensional protein, the structure approach will be used.

45. A query is a request for information sent to a database. Generally, queries are of three kinds: parameter-based (fill-in-the-blank-type searches), query-by-example (user-defined fields and values), and query languages (queries in a specific query language). A common query language used on the Internet is SQL, or structured query language, which makes use of "tables" and "select" statements to retrieve data from a database.

46. CGI is often used as part of Web sites to facilitate the dynamic communication between a user's computer ("client") and the computer on which the Web site resides ("server"). Web sites that have forms and menus for user input (e.g., e-mail address, name, platform) utilize CGI programs to process the data so that the server can act on it, either returning data based on the input (such as an updated, refreshed splash page) or further processing the input data (such as adding an e-mail address to a mailing list).

47. Although wet biological components such as the genome can be looked at metaphorically as "programs," the difference here is that in bioinformatics the genome is actually implemented as a database, and hence the intermingling of genetic and computer codes.

48. Francis Crick, "The Recent Excitement in the Coding Problem," *Progress in Nucleic Acids Research* 1 (1963): 163–217. See also Kay, *Who Wrote the Book of Life?*, 128–63.

49. For an example of this technique in action, see Celera's human genome report: Craig Venter et al., "The Sequence of the Human Genome," *Science* 291 (16 February 2001): 1304–51.

50. See Barbara Culliton, "Genomes, Proteomes, and Medicine," *Genome News Network* (12 February 2001): http://www.celera.com/genomics/news; William Clark, *The New Healers: The Promise and Problems of Molecular Medicine in the Twenty-First Century* (Oxford: Oxford University Press, 1999); and Christopher Mathew, "DNA Diagnostics: Goals and Challenges," *British Medical Bulletin* 55.2 (1999): 325–39.

51. See W. E. Evans et al., "Pharmacogenomics: Translating Functional Genomics into Rational

Therapeutics," *Science* 286 (15 October 1999): 487–91; Clark, *The New Healers*; Francis Collins, "Implications of the Human Genome Project for Medical Science," *JAMA* on-line 285.5 (7 February 2001): http://jama.ama-assn.org.

52. This is, broadly speaking, the approach of "regenerative medicine." See Sophie Petit-Zeman, "Regenerative Medicine," *Nature Biotechnology* 19 (March 2001): 201–6.

53. "Open source" has become an increasingly general term to signify any computer code that is freely available, and in the 1990s gained notoriety in part owing to the development of Linux, an open-source Unix-based operating system. For obvious reasons, software companies have, from the beginning, kept their source code under tight wraps, for the source code is, as its name implies, the infrastructure of the software, the very reason the software works at all. In characteristic "information-wants-to-be-free" terms, open-source movements have often been interpreted as offering a libratory, empowering access to computer technology in the face of corporate software culture. See Glyn Moody, *Rebel Code: Linux and the Spirit of the Information Age* (London: Penguin, 2001); Peter Ludlow, ed., *Crypto Anarchy, Cyberstates, and Pirate Utopias* (Cambridge: MIT Press, 2001).

54. Perl stands for Practical Extraction and Report Language and was developed by programmer Larry Wall in the 1980s, originally to handle text-processing needs that could be executed without needing to be compiled (they are instead interpreted). The Web site http://www.perldoc.com gives the following definition of Perl: "Perl is a high-level programming language with an eclectic heritage written by Larry Wall and a cast of thousands. It derives from the ubiquitous C programming language and to a lesser extent from sed, awk, the Unix shell, and at least a dozen other tools and languages. Perl's process, file, and text manipulation facilities make it particularly well-suited for tasks involving quick prototyping, system utilities, software tools, system management tasks, database access, graphical programming, networking, and world wide web programming. These strengths make it especially popular with system administrators and CGI script authors, but mathematicians, geneticists, journalists, and even managers also use Perl."

55. Perl's success in bioinformatics has spawned other more recent feats, which include the GigAssembler, a set of scripts written by James Kent at UCSD, which was instrumental in stitching together the sequence fragments of the Human Genome Project in time for the "big announcement" of its completion. See Lincoln Stein, "How PERL Saved the Human Genome Project," *Perl Journal:* http://www.tpi.com.

56. Bergson discusses the concepts of the virtual and the possible in several places, most notably in *Time and Free Will*, in which he advances an early formulation of "duration," and in *The Creative Mind*, in which the possible is questioned as being prior to the real. Although Bergson does not theorize difference in its poststructuralist vein, Deleuze's reading of Bergson teases out the distinctions that Bergson makes between the qualitative and quantitative in duration as laying the groundwork for a nonnegative (which, for Deleuze, means non-Hegelian) notion of difference as generative, positive, proliferative. See Gilles Deleuze, *Bergsonism* (New York: Zone Books, 1990), 91–103.

57. Deleuze's theory of difference owes much to Bergson's thoughts on duration, multiplicity, and the virtual. In *Bergsonism*, Delezue essentially recasts Bergson's major concepts (duration, memory, the "élan vital") along the lines of difference as positive, qualitative, intensive, and internally enabled.

58. As Deleuze states, "from a certain point of view, in fact, the possible is the opposite of the real, it is opposed to the real; but, in quite a different opposition, the virtual is opposed to the actual . . . The possible has no reality (although it may have an actuality); conversely, the virtual is not actual, but as such possesses a reality" (*Bergsonism*, 96).

59. A number of efforts are under way to assemble bioinformatics databases centered on these processes of regulation. BIND (Biomolecular Interaction Network Database) and ACS (Association for Cellular Signaling) are two recent examples. However, these databases must convert what is essentially time-based processes into discrete image or diagram files connected by hyperlinks, and, because the feasibility of dynamic databases is not an option for such endeavors, the results end up being similar to sequence databases.

60. As Susan Oyama has noted, there are a great deal of assumptions regarding the way in which the idea of biological data is configured; more often than not, genetic codes are assumed to remain

relatively static, while the environment is seen to be constantly changing. In this model, the organism is internally defined by a homeostasis in relation to a dynamic "outside." Similarly, the long-standing notions of matter and form take a new shape in the discourse of biotechnology and molecular genetics. The "logic of life" is seen to be innate to the genome, to DNA itself, as a primary matter (genotype) whose goal-directedness leads to the development of form (phenotype). As Oyama states, if information "is developmentally contingent in ways that are orderly but not preordained, and if its meaning is dependent on its actual functioning, then many of our ways of thinking about the phenomena of life must be altered" (*The Ontogeny of Information* [Durham, N.C.: Duke University Press, 2000], 3).

3. Wet Data

The heading "Remember, I need oxygen" is from the song "Remember I Was Vapour," by Gary Numan, from the album *Telekon* (Beggar's Banquet, 1980).

1. Greg Bear, *Blood Music* (New York: Ace, 1985), 60.
2. Ibid., 62.
3. Greg Bear, "Blood Music," in *Visions of Wonder,* ed. David Hartwell and Milton Wolf (New York: Tor, 1998), 47. This is the short-story version of the novel, which was subsequently expanded and included as a chapter in the novel.
4. Bear, *Blood Music,* 177.
5. National Research Council, *Microelectromechanical Systems: Advanced Materials and Fabrication Methods* (Washington, D.C.: National Academy Press, 1997). Also see the DARPA (Defense Advanced Research Projects Agency) MEMS program, at http://www.darpa.mil/MTO/MEMS; and Adrian Michalicek, "An Introduction to Microelectromechanical Systems," on-line presentation (2000): http://mems.colorado.edu/c1.res.ppt/ppt/g.tutorial/ppt.htm.
6. Antonio Regalado, "DNA Computing," *MIT Technology Review* (May/June 2000): http://www.techreview.com; Richard Lipton and Eric Baum, eds., *DNA Based Computers* (Princeton, N.J.: American Mathematical Society, 1996). DNA computing will be taken up in chapter 4.
7. Steven Levy, *Artificial Life* (New York: Vintage, 1992); Chris Langton, ed., *Artificial Life: An Overview* (Cambridge: MIT Press, 1995).
8. Michalicek, http://mems.colorado.edu/c1.res.ppt/ppt/g.tutorial/ppt.htm.
9. MEMS Clearinghouse: http://www.memsnet.org.
10. Ivan Amato, "May the Micro Force Be with You," *MIT Technology Review* (September/October 1999): http://www.technologyreview.com; Michael Ramsey, "The Burgeoning Power of the Shrinking Laboratory," *Nature Biotechnology* 17 (November 1999): 1061–62.
11. Roy Shuvo, Lisa Ferrara, Aaron Fleischman, and Edward Benzel, "Microelectromechanical Systems and Neurosurgery: A New Era in a New Millennium," *Neurosurgery* 49.4 (October 2001): 779–91; Andrew Marshall and John Hodgson, "DNA Chips: An Array of Possibilities," *Nature Biotechnology* 16 (January 1998): 27–31; Ann Caviani Pease et al., "Light-Generated Oligonucleotide Arrays for Rapid DNA Sequence Analysis," *Proceedings of the National Academy of Science* 91 (1994): 5022–26; Peter Mitchell, "Microfluidics: Downsizing the Large-Scale Laboratory," *Nature Biotechnology* 19 (August 2001): 717–21.
12. Glenn McGall et al., "Light-Directed Synthesis of High-Density Oligonucleotide Arrays Using Semiconductor Photoresists," *Proceedings of the National Academy of Science* 93 (1996): 13555–60.
13. A. Pedrocchi, S. Hoen, G. Ferrigno, and A. Pedotti, "Perspectives on MEMS in Bioengineering" *IEEE* 47.1 (January 2000): 8–11.
14. Shuvo et al., "Microelectromechanical Systems"; D. Williams, "The Right Time and the Right Place: The Concepts and Concerns of Drug Delivery Systems," *Medical Device Technology* 9.2 (March 1998): 10–12, 16; N. A. Polson and M. A. Hayes, "Microfluidics: Controlling Fluids in Small Places," *Analytical Chemistry* 73.11 (July 2001): 312–19A.

15. Williams, "The Right Time and the Right Place"; Pedrocchi et al., "Perspectives on MEMS"; W. Habib, R. Khankari, and J. Hontz, "Fast-Dissolve Drug Delivery Systems," *Critical Review of Therapeutic Drug Carrier Systems* 17.1 (2000): 61–72; M. Nakano, "Places of Emulsions in Drug Delivery," *Advances in Drug Delivery* 45.1 (6 December 2000): 1–4.

16. Ramsey, "The Burgeoning Power"; Mitchell, "Microfluidics"; Polson and Hayes, "Microfluidics"; J. Farinas, A. W. Chow, and H. G. Wada, "A Microfluidic Device for Measuring Cellular Membrane Potential," *Analytical Biochemistry* 295.2 (15 August 2001): 138–42.

17. P.A. Clarke, R. Poele, R. Wooster, and P. Workman, "Gene Expression Microarray Analysis in Cancer Biology, Pharmacology, and Drug Development: Progress and Potential," *Biochemical Pharmacology* 62.10 (2001): 1311–36; H. C. King and A. Sinha, "Gene Expression Profile Analysis by DNA Microarrays: Promise and Pitfalls," *JAMA* 286.18 (November 2001): 2280–88.

18. G. Wallraff et al., "DNA Sequencing on a Chip," *Chemtech* (February 1997): 22–32; Marshall and Hodgson, "DNA Chips"; Robert Service, "Microchip Arrays Put DNA on the Spot," *Science* 282 (16 October 1998): 396–401; S. Fodor, R. P. Rava, X. C. Huang, A. C. Pease, C. P. Holmes, and C. L. Adams, "Multiplexed Biochemical Assays with Biological Chips," *Nature* 364 (5 August 1993): 555–56; J. Madoz-Gurpide, H. Wang, D. E. Misek, F. Brichory, and S. M. Hanash, "Protein Based Microarrays: A Tool for Probing the Proteome of Cancer Cells and Tissues," *Proteomics* 1 (October 2001): 1279–87.

19. For details on microarray techniques, see Wallraff et al., "DNA Sequencing."

20. Lev Manovich, *The Language of New Media* (Cambridge: MIT Press, 2001), 46.

21. Williams, "The Right Time and the Right Place"; Shuvo et al., "Microelectromechanical Systems"; P. Norris, "MEMS: A Technological Solution to a Social Problem?" *Health Care Analysis* 6.4 (December 1998): 318–20.

22. See Marshall and Hodgson, "DNA Chips"; Karl Thiel, "The Matrix: A Revolution in Array Technologies," *Biospace.com* (7 June 1999): http://www.biospace.com. For more on microarrays, see the special issue of *Science* (7 April 1995).

23. Fred Sanger et al., "DNA Sequencing with Chain-Terminating Inhibitors," *Proceedings of the National Academy of Science* 74 (1977): 5463; Wallraff et al., "DNA Sequencing."

24. Fodor et al., "Multiplexed Biochemical Assays." This research, and other research like it, took its cue from microelectrical and materials engineering research into silicon substrates. See Kurt Petersen, "Silicon as a Structural Material," *IEEE Proceedings* (May 1982).

25. Pease et al., "Light-Generated Oligonucleotide Arrays."

26. The most notable formulations of the clockwork body come from Descartes and La Mettrie, both of whom view the body as simultaneously animalistic (that is, unthinking, nonconscious) and impersonally automated (hence the metaphor of the clock mechanism). See René Descartes, *Meditations on First Philosophy* (Cambridge: Cambridge University Press, 1993); and Julien Offray de La Mettrie, *Machine Man and Other Writings* (Cambridge: Cambridge University Press, 1996). For a critical-historical view of how the Cartesian clockwork body links up with the cybernetic, informatic body, see David Tomas, "Feedback and Cybernetics: Reimaging the Body in the Age of Cybernetics," *Cyberspace/Cyberbodies/Cyberpunk,* ed. Mike Featherstone and Roger Burrows (London: Sage, 1995), 21–45.

27. Catherine Waldby, *The Visible Human Project: Informatic Bodies and Posthuman Medicine* (New York: Routledge, 2000).

4. Biocomputing

The heading "So please just send in the machines" is from the song "Crash," by Gary Numan, from the album *Dance* (Beggar's Banquet, 1981).

1. Greg Egan, *Diaspora* (New York: HarperPrism, 1998), 210–51.
2. Ibid., 235.
3. Ibid., 242.
4. Ibid., 248–49.

5. For example, see William Aspray and Martin Campbell-Kelly, *Computer: A History of the Information Machine* (New York: Basic Books, 1996), 9–28.

6. The notion of "bioports" is a reference to David Cronenberg's film *eXistenZ*; the notion of the brain as computer memory is a reference to William Gibson's story "Johnny Mnemonic," in *Burning Chrome* (New York: Ace, 1983).

7. For more on bioinformatics, see chapter 2.

8. For popular versions of this claim, see Vincent Kiernan, "DNA-Based Computers Could Race Past Supercomputers, Researchers Predict," *Chronicle of Higher Education* (28 November 1997): http://chronicle.com. Such speculations are common among technovisionary theorists. See, for example, Ray Kurzweil, *The Age of Spiritual Machines* (New York: Penguin, 1999), 9–40.

9. For a range of proposed models for biocomputing, see the essays collected in Richard Lipton and Eric Baum, eds., *DNA Based Computers: DIMACS Workshop* (Princeton, N.J.: Princeton University Press, 1995). For science journalism on biocomputing, see Alan Dove, "From Bits to Bases: Computing with DNA," *Nature Biotechnology* 16 (September 1998): 830–32; and Antonio Regalado, "DNA Computing," *MIT Technology Review* (May/June 2000): http://www.techreview.com. For a technical book on biocomputing generally, see Christian Calude and Gheorghe Paun, *Computing with Cells and Atoms: An Introduction to Quantum, DNA, and Membrane Computing* (London: Taylor & Francis, 2001).

10. For a proof-of-concept article, see Leonard Adleman, "Molecular Computation of Solutions to Combinatorial Problems," *Science* 266 (11 November 1994): 1021–24. Much of the following analysis of biocomputing derives from Adleman's model of the DNA computer.

11. For applications in cryptography, see Dan Boneh, "Breaking DES Using a Molecular Computer," in Lipton and Baum, *DNA Based Computers*, 37–66. For applications in detection devices, see Ronald Breaker, "Engineered Allosteric Ribozymes as Biosensor Components," *Current Opinion in Biotechnology* 13 (2002): 31–39.

12. Most biocomputing research is being carried out in university labs. However, DARPA has had a biocomputing project for some time, and Bell Labs and Maxygen have funded biocomputing-related research in the past.

13. For more on various strategies in biocomputing, see Cristian Calude and Gheorghe Paun, "Computing with Cells and Atoms in a Nutshell," *Complexity* 6.1 (2001): 38–48; and Andrew Ellington, M. P. Robertson, K. D. James, and J. C. Cox, "Strategies for DNA Computing," in *DIMACS Series in Discrete Mathematics and Theoretical Computer Science,* vol. 48 (Princeton, N. J.: American Mathematical Society, 1999): 173–84.

14. See Adleman, "Molecular Computation of Solutions to Combinatorial Problems"; Leonard Adleman, "On Constructing a Molecular Computer," in Lipton and Baum, *DNA Based Computers*, 1–21; Martyn Amos, "DNA Computing," Ph.D. thesis, University of Warwick, Department of Computer Science, 1997; D. Boneh, C. Dunworth, R. Lipton, and J. Sgall, "On the Computational Power of DNA," personal Web site: http://www.cs.princeton.edu/~dabo; Ellington et al., "Strategies for DNA Computing"; Warren Smith, "DNA Computers *in vitro* and *vivo*," in Lipton and Baum, *DNA Based Computers*, 121–85.

15. See Calude and Paun, "Computing with Cells and Atoms in a Nutshell"; G. Berry and G. Bondol, "The Chemical Abstract Machine," *Theoretical Computer Science* 96 (1992): 217–48; G. Paun and J. Dassow, "On the Power of Membrane Computing," *Journal of Universal Computer Science* 5.2 (1999): 33–49; Gheorge Paun, "Computing with Membranes: An Introduction," *Bulletin of the EATCS* 68 (1999): 139–52.

16. See Martyn Amos and Gerald Owenson, "Cellular Computing," *ERCIM News* 43 (October 2000): http://www.ercim.org; Masami Hagiya, "From Molecular Computing to Molecular Programming," in *DNA Computing 2000*, ed. A. Condon (Berlin: Springer, 2001); Gheorge Paun, "Computing by Splicing," *Theoretical Computer Science* 168 (1996): 321–66.

17. For speculations on biological computers, see C. H. Bennett, "Logical Reversibility of Computation," *IBM Journal of Research and Development* 17 (1973): 525–32. Bennett mentions the similarities between gene expression and the Turing machine.

18. For an introduction to graph theory, see Gary Chartrand, *Introductory Graph Theory* (New York: Dover, 1977); and Robin Wilson, *Introduction to Graph Theory* (New York: Addison-Wesley, 1997). For a historical overview, see Norman Biggs, E. Keith Lloyd, and Robin J. Wilson, *Graph Theory, 1736–1936* (Oxford: Clarendon, 1976). In addition, most college-level textbooks in discrete mathematics will contain chapters on graphs, trees, maps, and subjects in topology.

19. See Chartrand, *Introductory Graph Theory*, 67–76.

20. For an introductory explanation of PCR, see Susan Barnum, *Biotechnology: An Introduction* (New York: Wadsworth, 1998), 61–63. For an anthropological-sociological analysis of PCR as a tool of biotechnology, see Paul Rabinow, *Making PCR: A Story of Biotechnology* (Chicago: University of Chicago Press, 1996).

21. For an explanation of electrophoresis, see Barnum, *Biotechnology*, 62–64.

22. The discursive relationships between cybernetics and molecular biology have been explored at length elsewhere. Lily Kay's *Who Wrote the Book of Life?* (Stanford, Calif.: Stanford University Press, 2000) remains the most comprehensive and most intelligent tracing of this complex history, while Evelyn Fox Keller's *Refiguring Life: Metaphors of Twentieth-Century Biology* (New York: Columbia University Press, 1995) and Richard Doyle's *On Beyond Living* (Stanford, Calif.: Stanford University Press, 1997) have explored the implications of this relationship in relation to current theories of language, media, and materiality.

23. For a key research paper, see François Jacob and Jacques Monod, "Genetic Regulatory Mechanisms in the Synthesis of Proteins," *Journal of Molecular Biology* 3.318 (1961): 318–59. For an extended analysis, see Kay, *Who Wrote the Book of Life?*, 193–234.

24. The full citation is Alan Turing, "On Computable Numbers with an Application to the *Entscheidungsproblem*," *Proceedings of the London Mathematical Society* 2.42 (1936): 230–67. This chapter does not claim to provide analyses of the mathematics in Turing's papers, but rather takes up their general claims as they relate to both computer science and biocomputing at the philosophical level.

25. For more on Gödel, see John Casti and Werner Depauli, *Gödel: A Life of Logic* (New York: Perseus, 2001).

26. Turing's computability thesis applied to biocomputing asks how a biological system can validate its own "statements" using only its internal set of procedures and interactions. In short, "computability" for biological computer systems is akin to an inquiry into the consistency and accuracy of molecular immunology and genetic regulation. We can take these two examples as instances in molecular biology (immunology and genetics) in which the body, at the biomolecular level, performs a kind of "error checking" as part of its normal functioning. In molecular immunology, it has been known for some time that the immune system displays a highly refined level of molecular specificity in the "recognition" of foreign agents (or "antigens"). The immune system produces killer T cells with a highly specific structure that literally binds or docks to the foreign molecule, thereby dismantling it. In a like fashion, the genetic process of protein synthesis (the "translation" of DNA into RNA, and the "transcription" of RNA into a protein) has been shown to display a sophisticated error-detection apparatus by using RNA to perform "editing" after the transcription phase has happened. Both instances—in molecular immunology and genetic protein synthesis—are examples of modules within the system that perform a validation of the system's functioning itself. In Turing's terms, the "statements" of these biological systems ("apply antibody X to antigen Y"; "check protein X against gene Y") are validated structurally and via biochemical interactions.

27. For examples, see Kurzweil, *The Age of Spiritual Machines*; Marvin Minsky, *Society of Mind* (New York: Simon and Schuster, 1985); Hans Moravec, *Robot: Mere Machine to Transcendent Mind* (Oxford: Cambridge University Press, 1999).

28. As is well documented, the U.S. military funded a number of research programs during and after World War II, aimed at the development of the burgeoning field of computer engineering for military, security, and communications application. Von Neumann was often hired as a scientific adviser on such projects, including those sponsored by the National Defense Research Committee's Ballistics Research Laboratory (BRL), whose main interest was in using computers to efficiently and accurately calculate firing tables for a range of artillery. Arising from purely practical concerns, the

von Neumann architecture attempted to address the limited storage capacity of the computers built up until that time. Inspired by the "delay-line storage" developed by J. Presper Eckert for the ENIAC, von Neumann and the EDVAC project team came to the realization that the computer's storage device could hold both the program instructions and the data on which that program would operate. See William Aspray, *John von Neumann and the Origins of Modern Computing* (Cambridge: MIT Press, 1990).

29. John von Neumann, *The Computer and the Brain* (New Haven: Yale University Press, 2000), 29.

30. Ibid., 68.

31. Ibid., 66.

32. See Alan Hodges, *Alan Turing: The Enigma* (New York: Simon and Schuster, 1983).

33. See Alan Turing, "Computing Machinery and Intelligence," *Mind* 49.236 (October 1950): 433–60.

34. Turing illustrates this with a simple example and a thought experiment. The simple example is the notion of a chess-playing computer. Someone has indeed programmed the computer to play chess according to the finite rules of chess, and the chess-playing computer simply carries out a series of instructions, but the precise outcome or prediction of moves in any given game is predicated on the human opponent's moves.

35. The Turing test has particular relevance for telecommunications technologies such as the Internet, where a host of software applications—IRC, MUDs, IM, and, of course, e-mail—seem to be Turing tests put into practice. The discourse surrounding "virtual cross-dressing" and negotiations of race, gender, and sexuality on-line would seem to be extensions of Turing's original "imitation game," in which gender and cultural assumptions are at play.

36. It is noteworthy that Turing's paper begins by describing the "imitation game" as one between human players only: "It is played with three people, a man (A), a woman (B), and an interrogator (C) who may be of either sex. The interrogator stays in a room apart from the other two. The object of the game for the interrogator is to determine which of the other two is the man and which is the woman." A number of media theorists, such as Katherine Hayles and Alison Adams, have pointed to the ways in which Turing frames a computer-science question in terms of gender and sexuality. We might also pose the following question: if the Turing test points to the performative dimensions (that is, the social and cultural dimensions) of intelligence, in relation to biocomputers a Turing test might similarly point to the performative dimensions of biological "life." Life would be that which behaves as life, or that adequately performs life. See Alison Adams, *Artificial Knowing: Gender and the Thinking Machine* (New York: Routledge, 1998); and Katherine Hayles, *How We Became Posthuman* (Chicago: University of Chicago Press, 1999).

37. Turing, "Computing Machinery and Intelligence," 435.

38. It should be noted that Turing's late work in the mathematical foundations of morphogenesis is an important exception. It could be said that Turing's interest in morphogenesis is a continuation of his interest in adaptability, development, and growth in biological or computational systems. As a mathematician, Turing's interests are similar to, but also different from, that of James Watson, Francis Crick, Rosalind Franklin, and other molecular biologists working at the same time on DNA and genetic material. Even though Turing approaches morphogenesis from the perspective of mathematics, he still does not conceive of morphogenesis in specifically computational terms. By contrast, molecular biologists such as Francis Crick, François Jacob, and Jacques Monod import the concepts and language of cybernetics into their research into gene regulation, and move closer to a notion of the cell as a computer, though this is still at the level of metaphor.

39. What should be apparent in these two paradigms of thinking about humans and computers is both the anthropocentrism and the emphasis on defining the organism in terms of cognitive processes. Both Turing and von Neumann are only concerned about the body to the extent that it provides a framework or hardware on top of which higher-level processes can occur. Even von Neumann's materialist approach, focusing as it does on developments in neuroscience, displays a predilection to consider the nervous system solely in terms of brain activity, while Turing's test appears to raise the question of embodiment, only to abstract it behind the logical operators seen to inhere in language. In these examples of computer science, we can detect more than a hint of biologism: the Turing test

assumes as its level of success the intractability of gender from biological sex, while the von Neumann architecture employs a constructionist view of cognition as proceeding from aggregates of lower-level functions in neurons or switches. This biologism provides the foundation for further thinking about computers as more than computers; calculation and "computability" provide the materialist correlatives for the constructionist view of cognition. To compute, therefore, becomes the functional analogue of intelligence (in terms of learning) and memory (in terms of data storage), beneath which runs the lower-level hardware or biology of the system.

40. The slogan of mainframe computing is "never mind that man behind the curtain"; the slogan of biocomputing is "even cells do it."

41. See Gilles Deleuze and Félix Guattari, *A Thousand Plateaus: Capitalism and Schizophrenia*, trans. Brian Massumi (Minneapolis: University of Minnesota Press, 1987), 45–48, 57–60, 233–309. The concept of "the molecular" often appears under different guises, in discussions on multiplicity, stratification, or deterritorialization.

42. In this, Deleuze and Guattari follow Henri Bergson's claim that the continuous, qualitative change of duration that characterizes living organisms occurs through becoming. For Bergson—as for Deleuze and Guattari—becoming is a subject, not a verb, a line, not a point. Bergson points to the examples of physiology/movement, embryogenesis, and morphogenesis to suggest that we artificially designate static points along the trajectory of a movement or development, culminating in an image of becoming that is discrete change (serial, sequential, stop/go). Bergson's interest lies less in this quantitative method than in an "intuitive" one that belongs to philosophy, and that involves beginning from continuous becoming, rather than from discrete states. The key to this shift—for Bergson as for Deleuze's work influenced by Bergson—is to take dynamic change or "duration" as a starting point. See Deleuze and Guattari's discussion of segmentarity in *A Thousand Plateaus*, 202–26. Also see Gilles Deleuze, *Bergsonism* (New York: Zone, 1991).

43. Although graph theory accommodates situations in which nodes can be edges (nested graphs/networks), what is important to note is that such accommodations require a shift in scale, such that the node-edge topology can be fulfilled. This node-edge topology enables a set of quantitative analyses of networks conceived as such (degrees of nodes, network dimensions, adjacency matrices, and so forth).

44. This is illustrated in contemporary research that aims to map biomolecular interactions in databases, such as WIT (What Is There?; at http://wit.mcs.anl.gov/WIT2) and BIND (Biomolecular Interaction Network Database; at http://www.bind.ca).

5. Nanomedicine

The heading "Here we are, we drift like gas" is from the song "A Subway Called You," by Gary Numan, from the album *Dance* (Beggar's Banquet, 1981).

1. Linda Nagata, *The Bohr Maker* (New York: Bantam, 1995).

2. Other, like-minded SF works that deal with nanotechnology include Greg Bear, / *[Slant]* (New York: Tor, 1997); Neal Stephenson, *The Diamond Age* (New York: Bantam, 1995); Kathleen Ann Goonan, *Queen City Jazz* (New York: Tor, 1994); and Jack Dann and Gardner Dozois, eds., *Nanotech* (New York: Ace, 1998).

3. Eric Drexler, *Engines of Creation: The Coming Era of Nanotechnology* (New York: Doubleday, 1986), 172–73.

4. This move, from transcendence to immanence, is demonstrated in *The Bohr Maker* and other nanotech SF by, on the one hand, a proliferating, generative, perhaps epidemic spread of a new type of matter, and, on the other hand, by the imagined transformations in consciousness and life caused by this spread. These constitute "ambivalent" miasmas in that the SF authors seem to want to point to a new type of elevated consciousness, while also wanting to point to a radically new view of what biological life or matter itself can be when transformed by nanotechnologies. See also Bear, / *[Slant]*, and the stories collected in Dann and Dozois, *Nanotech,* for more examples.

5. Richard Feynman, "There's Plenty of Room at the Bottom," in *Nanotechnology: Research and Perspectives*, ed. B. C. Crandall and James Lewis (Cambridge: MIT Press, 1992); also available on-line at http://www.zyvex.com/nanotech/feynman.html.

6. Drexler, *Engines of Creation*, 4–5.

7. Eric Drexler, Chris Peterson, and Gayle Pergamit, *Unbounding the Future: The Nanotechnology Revolution* (New York: William Morrow, 1991), 5. Also available on-line at http://www.foresight.org.

8. Drexler, *Engines of Creation*, 14–23.

9. Robert Freitas, "Say 'Ah' for Nanomedicine," *The Sciences* (July/August 2000): 26–31.

10. Robert Freitas, "Exploratory Design in Medical Nanotechnology: A Mechanical Artificial Red Cell," *Artificial Cells, Blood Substitutes, and Biotechnology* 26 (1998): 411–30. Updated version published as "Respirocytes: A Mechanical Artificial Red Cell," available on-line at http://www.foresight.org/Nanomedicine/Respirocytes.html.

11. For an overview, see David Voss, "Nanomedicine Nears the Clinic," *MIT Technology Review* (January/February 2000): http://www.techreview.com; Michael Gross, *Travels to the Nanoworld: Miniature Machinery in Nature and Technology* (New York: Plenum, 1999); and the special issue "Nanotech: The Science of the Small Gets Down to Business," *Scientific American* (September 2001).

12. Jeffrey Soreff, "Recent Progress: Steps toward Nanotechnology," *IMM Report* 19 (October 2000): http://www.imm.org; Colin Macilwain, "Nanotech Thinks Big," *Nature* 405 (15 June 2000): 730–32.

13. National Science and Technology Council, "National Nanotechnology Initiative: Leading the Way to the Next Industrial Revolution," NNI Web site: http://www.nanotech.gov.

14. Eric Drexler, "Protein Design as a Pathway to Molecular Manufacturing," *Proceedings of the National Academy of Science* 78 (September 1981): 5275–78.

15. Ibid., 1.

16. Drexler, Peterson, and Pergamit, *Unbounding the Future*, chapter 1, http://www.foresight.org/UTF/Unbound_LBW.

17. Ibid.

18. Robert Freitas, *Nanomedicine*, vol. 1, *Basic Capabilities* (New York: Landes Bioscience, 1999), 2. Also at http://www.nanomedicine.com.

19. Ibid., 25.

20. Drexler, *Engines of Creation*, 99.

21. Freitas, *Nanomedicine*, chapter 1, http://www.nanomedicine.com.

22. Drexler, Peterson, and Pergamit, *Unbounding the Future*, chapter 10, http://www.foresight.org/UTF/Unbound_LBW.

23. Freitas, *Nanomedicine*, chapter 1, http://www.nanomedicine.com.

24. Drexler, *Engines of Creation*; B. C. Crandall, ed., *Nanotechnology: Molecular Speculations on Global Abundance* (Cambridge: MIT Press, 1999); Ralph Merkle, "It's a Small, Small, Small, Small World," *MIT Technology Review* (February/March 1997), also available on-line: http://www.zyvex.com.

25. Drexler, Peterson, and Pergamit, *Unbounding the Future*, chapter 6, http://www.foresight.org/UTF/Unbound_LBW.

26. Freitas, "Respirocytes," Abstract, http://www.foresight.org/Nanomedicine/Respirocytes.html. All citations from this article are from the updated, on-line version of the paper.

27. Ibid., section 2.

28. Ibid., section 2.2.2 ("Molecular Sorting Rotors") and section 2.2.3 ("Sorting Rotor Binding Sites").

29. Ibid., section 6.

30. Robert Freitas, "Nanomedicine: Is Diamond Biocompatible with Living Cells?" *IMM Report* 12 (December 1999): http://www.imm.org.

31. Freitas, "Respirocytes," section 6, http://www.foresight.org/Nanomedicine/Respirocytes.html.

32. Soreff, "Recent Progress"; B. Yurke, A. J. Tuberfield, A. P. Mills, F. C. Simmel, and J. L. Neumann, "A DNA-Fuelled Molecular Machine Made of DNA" *Nature* 406 (10 August 2000): 605–8; George Banchand, P. K. Soong, H. P. Neves, A. G. Olkhavets, H. G. Graighead, and C. D. Monte-

magno, "Powering an Inorganic Nanodevice with a Biomolecular Motor," *Science* 290 (24 November 2000): 1555–58.

33. James Watson and Francis Crick, "General Implications of the Structure of Deoxyribonucleic Acid," *Nature* 171 (1953): 964–67.

34. Drexler, *Engines of Creation*, 15, 58–63.

35. Cited in "DNA 'Motors' for Computer Processing Create a Stir," Bell Labs Press Release (August 2000): http://www.bell-labs.com.

36. Fredric Jameson, *Postmodernism, or, the Cultural Logic of Late Capitalism* (Durham, N. C.: Duke University Press, 1995); Jean Baudrillard, *Simulations* (New York: Semiotext[e], 1983); Steven Best and Douglas Kellner, *The Postmodern Turn* (New York: Guilford, 1997); Katherine Hayles, *How We Became Posthuman* (Chicago: University of Chicago Press, 1999).

37. Richard Doyle, *On Beyond Living: Rhetorical Transformations of the Life Sciences* (Stanford, Calif.: Stanford University Press, 1997), 11.

38. Ibid.

39. Lily Kay, "Who Wrote the Book of Life? Information and the Transformation of Molecular Biology, 1945–55," *Science in Context* 8.4 (1995): 609–34; Evelyn Fox Keller, *Refiguring Life: Metaphors of Twentieth-Century Biology* (New York: Columbia University Press, 1995).

40. Doyle, *On Beyond Living*, 13.

41. Ibid., 20.

42. Sean Cubitt, "Supernatural Futures: Theses on Digital Aesthetics," in *Future Natural*, ed. George Robertson et al. (London: Routledge, 1996).

43. Drexler, *Engines of Creation*, 99–117.

44. Donna Haraway, *Simians, Cyborgs, and Women: The Reinvention of Nature* (New York: Routledge, 1991), 164.

45. Hayles, *How We Became Posthuman*, 1–18.

46. Drexler, *Engines of Creation*, 58–63.

47. Hayles, *How We Became Posthuman*, 50–70.

6. Systems Biology

The heading "Where is my outline—I start to fade" is from the song "Telekon," by Gary Numan, from the album *Telekon* (Beggar's Banquet, 1980).

1. Bruce Sterling, "Swarm," *Schismatrix Plus* (New York: Ace, 1996), 246. For more on the social, political, and cultural effects of colonialism, see Bill Ashcroft, Gareth Griffiths, and Helen Tiflin, eds., *The Post-Colonial Studies Reader* (New York: Routledge, 1995).

2. Sterling, "Swarm," 255.

3. Ibid., 256.

4. In the author's preface, Sterling acknowledges some debt to the theories of chaos and complexity, and indeed biology's study of insect colonies has for some time been an inspiration to the sciences of complexity, spawning an offshoot of research in computer science and artificial life studying simulated swarming and flocking behavior. What is interesting to note in "Swarm" is that the author has difficulty conceiving of a purely distributed living system: the introduction of the fleshy Swarm intelligence at times ends up recuperating the immanent network logic of the Swarm as a whole. It is, curiously, as if the author introduced an AI-based notion of "mind" into a distributed living system, thereby reestablishing Cartesian dualism, if only temporarily. Because Sterling is a writer who always attempts to pay attention to the politics of his fictions, this tension is duplicated on the political level, between the naive naturalism of the communal Swarm (where there is no privacy, no sense of property, though a lot of exchange of biomaterials), and the hierarchical order temporarily established by the Swarm intelligence, Queen, and worker-bee symbionts.

5. It should be stated that this chapter will focus on the relationship between "net and self" (to paraphrase Manuel Castells) within molecular biology and biotechnology. However, the tension is

already evident in other life-science fields such as anatomy and medical practices such as telesurgery. A glance at a modern anatomical textbook reveals this tension. Most textbooks contain an introductory chapter that proceeds to dissect the human body, either moving inside out (from DNA to the skeletal or muscular system) or moving outside in (from the whole body to molecules). The constructionist logic of anatomy (wholes/parts) gradually gives way to a more informatic logic of molecular genetics (codes). However, this seemingly neat transition is also frustrated by anatomy itself, which has, since Vesalius, been preoccupied with the body as a collection of fragments and parts, which could be dis-played, diagrammed, and inserted into tables in the anatomical texts of the early-modern period. For more on the nonanthropomorphic nature of anatomy, see Jonathan Sawday's historical approach in *The Body Emblazoned* (New York: Routledge, 1995), as well as Catherine Waldby's technoscientific approach in *The Visible Human Project* (London: Routledge, 2000).

6. See Leroy Hood, "The Human Genome Project and the Future of Biology," *Biospace.com* (1999): http://www.biospace.com. Hood is one of the most prominent supporters of the systems biology approach. As one of the members of the original planning committee for the Human Genome Project in the late 1980s, Hood went on to develop the first gene sequencing machines at CalTech, and founded a company, Applied Biosystems, to market the technology. After serving as chair of the molecular biology department at the University of Washington, Hood went on to form the Institute for Systems Biology, an independent, nonprofit research group. See "Under Biology's Hood: Q & A with Leroy Hood," *MIT Technology Review* (September 2001): http://www.technologyreview.com.

7. Ludwig von Bertalanffy, generally credited with developing systems theory, addressing this paradox in the 1969 preface to the revised edition of his *General System Theory*, notes how, on the one hand, systems theory had, by the time of his writing, established journals, courses, textbooks, and colloquiums. On the other hand, Bertalanffy also stressed the need to remain articulate and specific about systems theory in its scientific approaches. See Ludwig von Bertalanffy, *General System Theory* (New York: George Braziller, 1989), xvii–xxiv.

8. Ibid., 3–10.

9. Humberto Maturana and Francesco Varela, *The Tree of Knowledge* (Boston: Shambhala, 1998), 216.

10. Bertalanffy, *General System Theory*, 17–29; Francesco Varela, "The Emergent Self," in *The Third Culture*, ed. John Brockman (New York: Touchstone, 1995), 211–12.

11. For better pop-science accounts and introductions to the sciences of complexity, see Fritjof Capra, *The Web of Life: A New Understanding of Living Systems* (New York: Doubleday, 1997); Brockman, *The Third Culture*; Stuart Kauffman, *At Home in the Universe: The Search for the Laws of Self-Organization and Complexity* (Oxford: Oxford University Press, 1995); Mitchell Waldrop, *Complexity: The Emerging Science at the Edge of Order and Chaos* (New York: Simon and Schuster, 1992); John Casti, *Complexification* (New York: HarperPerennial, 1995); John Holland, *Hidden Order: How Adaptation Builds Complexity* (Reading, Mass.: Perseus, 1995); Steven Johnson, *Emergence: The Connected Lives of Ants, Brains, Cities, and Software* (New York: Scribner's, 2001); and James Gleick, *Chaos: Making a New Science* (New York: Penguin, 1988). Unsurprisingly, such introductions tend more often than not toward the "gee-whiz" end of the spectrum, something that is somewhat of a hallmark of the pop-science book genre.

12. See Johnson, *Emergence*, 117–23; and Terry Bossomaier and David Green, *Patterns in the Sand: Computers, Complexity, and Everyday Life* (Reading, Mass.: Perseus, 1998), 159–75.

13. See Kauffman, *At Home in the Universe*, 25–28, 273–304.

14. Bertalanffy, *General System Theory*, 32.

15. See Lily Kay, *Who Wrote the Book of Life? A History of the Genetic Code* (Stanford, Calif.: Stanford University Press, 2000); Evelyn Fox Keller, *Refiguring Life: Metaphors of Twentieth-Century Biology* (New York: Columbia University Press, 1995); and Steve Heims, *Constructing a Social Science for Postwar America: The Cybernetics Group 1946–1953* (Cambridge: MIT Press, 1993).

16. Claude Shannon and Warren Weaver, *The Mathematical Theory of Communication* (Chicago: University of Illinois Press, 1965), 3–10, 32–35.

17. Norbert Wiener, *Cybernetics: Or Control and Communication in the Animal and the Machine* (Cambridge: MIT Press, 1996), 6–12, 60–64.

18. Ibid., 95–98.

19. Norbert Wiener, *The Human Use of Human Beings: Cybernetics and Society* (New York: Da Capo, 1954), 15–27, 37–38, 95–96.

20. Lily Kay, "Cynernetics, Information, Life: The Emergence of Scriptural Representations of Heredity," *Configurations* 5.1 (1997): 23–91.

21. Francis Crick, "The Genetic Code," *Scientific American* 207 (1962): 66–75; "The Genetic Code—Yesterday, Today, and Tomorrow," *Cold Spring Harbor Symposia on Quantitative Biology* 31 (1966): 3–9; "The Present Condition of the Coding Problem," *Brookhaven National Laboratory Symposia* (June 1959): 35–39. See Kay, *Who Wrote the Book of Life?*, 141–63.

22. Keller, *Refiguring Life*, 89–93.

23. Bertalanffy, *General System Theory*, 6.

24. Ibid., 37.

25. Ibid., 55. This observation—that complex global behavior may arise from determined local interactions—is one of the oft-repeated themes in the sciences of complexity. The challenge to complexity as a science was to move beyond abstract description and to develop modes of modeling this behavior.

26. Ibid., 186–94; Ludwig von Bertalanffy, *Perspectives on General System Theory* (New York: George Braziller, 1967), 97–101.

27. Bertalanffy, *General System Theory*, 39–46.

28. Ibid., 145–53.

29. Bertalanffy uses the following equation to begin his discussion of systems: $dQ_{n/dt} = f_n (Q_1, Q_2, \ldots Q_n)$. As their name indicates, differential equations calculate the quantitative difference in selected variables of a system under observation. The state of the system at a given moment (Q) is therefore a function (f) of the changes in particular variables of that system over time ($Q_1, Q_2, \ldots Q_n$). In short, because a system not only has multiple parts, but parts that change in time, any change to a variable in the system will have, theoretically, repercussions (however great or small) throughout the system as a whole. See ibid., 54–60.

30. Ibid., 55.

31. Bertalanffy states, concerning constitutive characteristics, that "constitutive characteristics are those which are dependent on the specific relations within the complex; for understanding such characteristics we therefore must know not only the parts, but also the relations" (ibid.). This approach in systems theory therefore does not designate elements or their groupings outside of the systemic context. In the case of molecular genetics, a summative grouping would be either a quantitative grouping of genes (number of genes per species) or a grouping of genes based on their isolated function (a database of protein-coding genes, regulatory genes, SNPs, etc.). Both of these are, in fact, common ways of organizing genetic information in genomics and bioinformatics. By contrast, a constitutive approach would interlink studies of composition, structure, and function with participation in various biopathways (metabolism, signal processing, storage). Therefore, a database of pathways may contain what would seem to be heterogeneous elements (coding and noncoding DNA, RNA, amino acids, enzymes), but their grouping is not done in isolation, but as part of a constitutive process.

32. Ibid., 141.

33. Ibid., 120–31.

34. Ibid., 57–59.

35. Bertalanffy describes equifinality as "the fact that the same final state can be reached from different initial conditions and in different ways" (ibid., 79).

36. Ibid., 68.

37. Ibid., 39.

38. Ibid., 158.

39. Ibid., 140–41.

40. Ibid., 163.

41. However, if, as we have seen, Bertalanffy makes the link between open systems and the steady state the distinguishing factor between living and nonliving systems, this also begs the question of the difference between living systems and open systems; that is, are all living systems also open systems for Bertalanffy? If so, a "general systems theory" would not be universal, as claimed, but would rather be quite specific to the biological domain. The issue is worth noting because of a methodological tension in Bertalanffy's systems theory. Although he begins emphasizing "relations" over "components," "parts," or things, he ends up reestablishing the division between living and nonliving based on substance or components (organisms versus machines). Theoretically, the relations-based approach would make no distinction between the substance or components of a system, as long as it fulfilled the central criterion of functioning as an open system through a steady state. Although Bertalanffy admits that mechanistic, cybernetic models are relevant to "secondary," autonomic subsystems, in the end he holds fast to the ontological division between living and nonliving, organism and machine. We can locate a tension in Bertalanffy's articulation of living systems here, which is a tension between pattern and matter—a tension that cybernetics and genetics articulate as one between materiality and data, genetic and computer codes.

42. Bertalanffy, *General System Theory*, 132.

43. Ibid., 140.

44. Ibid., 139.

45. Ibid., 39–48.

46. T. Ideker, V. Thorsson, J. A. Ranish, R. Christmas, J. Buhler, J. K. Eng, R. Baumgarner, and D. R. Goodlett, "Integrated Genomic and Proteomic Analyses of a Systematically Perturbed Metabolic Network," *Science* 292 (4 May 2001): 929–34. For other examples of like-minded work, see B. Schwikowski, P. Uetz, and Stanley Fields, "A Network of Protein-Protein Interactions in Yeast," *Nature Biotechnology* 18 (December 2000): 1257–61; H. Jeong, B. Tombor, R. Albert, Z. N. Olfvair, and A. L. Barabasi, "The Large-Scale Organization of Metabolic Networks," *Nature* 407 (5 October 2000): 651–54; and T. Norman, D. L. Smith, P. K. Sorger, B. L. Drees, S. M. O'Rourke, T. R. Hughes, J. C. Roberts, S. H. Friend, S. Fields, and A. W. Murray, "Genetic Selection of Peptide Inhibitors of Biological Pathways," *Science* 285 (23 July 1999): 591–95.

47. Kay, *Who Wrote the Book of Life?*, 150–55, 212–22.

48. Sociologist Manuel Castells, in *The Rise of the Network Society* (London: Blackwell, 1996), describes the effects of global capitalism, on the one hand, and the new social movements, on the other, as a tension between "net" and "self": "In a world of global flows of wealth, power, and images, the search for identity, collective or individual, ascribed or constructed, becomes the fundamental source of social meaning" (3).

49. See Donna Haraway, *Modest.Witness@Second_Millennium* (New York: Routledge, 1997); Richard Doyle, *On Beyond Living: Rhetorical Transformations of the Life Sciences* (Stanford, Calif.: Stanford, University Press, 1997); Dorothy Nelkin and Susan Lindee, *The DNA Mystique: The Gene as a Cultural Icon* (New York: W. H. Freeman, 1995); and Ruth Hubbard and Elijah Wald, *Exploding the Gene Myth* (Boston: Beacon, 1997).

50. For more on the development of early-modern anatomy, anatomical norms, and dissection, see Jonathan Sawday, *The Body Emblazoned: Dissection and the Human Body in Renaissance Culture* (New York: Routledge, 1995); and Nancy Siraisi, *Medieval and Early Renaissance Medicine* (Chicago: University of Chicago Press, 1990). Sawday's text is particularly useful in articulating how the technique of anatomical dissection, with its use of anatomy theaters and anatomical texts (replete with tables, lists, and diagrammatic logic), contributes to a means of materializing the modern-scientific view of the body as expressly anatomical (the whole composed of parts).

51. Bertalanffy, *General System Theory*, 52.

52. See Kay, *Who Wrote the Book of Life?*, 246–56.

53. The operon model was developed by Jacob and Monod in their well-known research on gene

expression and the role of "messenger RNA" in this process. They showed how, in a self-referential fashion, DNA produced proteins, whose sole role was to act as "repressors" or "promoters" back on DNA itself, blocking or facilitating the transcription of DNA into RNA. Such genes, which coded for proteins whose sole purpose was to regulate the production of genes, they called "regulatory genes." See François Jacob and Jacques Monod, "Genetic Regulatory Mechanisms in the Synthesis of Proteins," *Journal of Molecular Biology* 3 (1961): 318–59; and Kay, *Who Wrote the Book of Life?*, 222–28.

54. That being said, systems biology is still ambivalent as to whether it treats the issues of hierarchy in systems the same as medical genetics does. It must deal theoretically with issues such as hierarchy within systems, systemic closure, and autonomy. Depending on how it responds to these issues, it may either push to transform our understanding of the biomolecular body or reinforce the dominant paradigm of the biotech industry and medical genetics.

55. See Stuart Kauffman, *The Origins of Order: Self-Organization and Selection in Evolution* (New York: Oxford University Press, 1993); "Gene Regulation Networks: A Theory for Their Global Structure and Behavior," *Current Topics in Developmental Biology* 6 (1971): 145; "The Large-Scale Structure and Dynamics of Gene Control Circuits: An Ensemble Approach," *Journal of Theoretical Biology* 44 (1974): 167; and "New Questions in Genetics and Evolution," *Cladistics* 1 (1985): 247.

56. Kauffman, *At Home in the Universe*, 49.

57. Ibid., 48.

58. Ibid., 50.

59. Jacob and Monod, "Genetic Regulatory Mechanisms."

60. H. H. Goldstein, *The Computer from Pascal to von Neumann* (Princeton, N.J.: Princeton University Press, 1972).

61. Kauffman, *At Home in the Universe*, 73.

62. Ibid., 78. The concept of attractors is borrowed from its use in nonlinear dynamics and chaos theory. Kauffman, however, uses it in a particular way, with regard to sufficiently complex biological systems.

63. Kauffman finds that even when K is much lower than N, say K = 4, the number of possible states is hyperastronomical; however, K = 2 settles into a relatively stable state cycle. Kauffman also finds that, along with N = K equations, turning the control parameters, or P bias, can prompt hyperstable networks to become chaotic, and vice versa.

64. Kauffman, *At Home in the Universe*, 8.

65. See Maturana and Varela, *The Tree of Knowledge*; Humberto Maturana and Francesco Varela, *Autopoiesis and Cognition: The Realization of the Living* (Boston: D. Reidel, 1972); Niklas Luhmann, *Social Systems* (Stanford, Calif.: Stanford University Press, 1995); Milan Zeleny, ed., *Autopoiesis: A Theory of Living Organizations* (New York: Columbia University Press, 1981); William Rasch and Cary Wolfe, eds., *Observing Complexity: Systems Theory and Postmodernity* (Minneapolis: University of Minnesota Press, 2000); and Francesco Varela, Evan Thompson, and Eleanor Rosch, *The Embodied Mind* (Cambridge: MIT Press, 2000).

66. Maturana and Varela, *The Tree of Knowledge*, 34.

67. Ibid., 23.

68. For more on the observer–system relation, see Heinz von Foerster, *Observing Systems* (Seaside, Calif.: Intersystems, 1981).

69. Maturana and Varela, *Autopoiesis and Cognition*, 76.

70. Ibid., 79.

71. Ibid., 92.

72. Ibid.

73. Ibid.

74. Maturana and Varela, *The Tree of Knowledge*, 47.

75. As they state in *The Tree of Knowledge*: "That living beings have an organization, of course, is proper not only to them but also to everything we can analyze as a system. What is distinctive about them, however, is that their organization is such that their only product is themselves, with no sepa-

ration between producer and product. The being and doing of an autopoietic unity are inseparable, and this is their specific mode of organization" (48–49).

76. Ibid., 69.

77. Maturana and Varela, *Autopoiesis and Cognition*, 98.

Conclusion

The heading "Just pretend it's for real" is from the song "Slowcar to China," by Gary Numan, from the album *Dance* (Beggar's Banquet, 1981).

1. Up until recently, the main presidential bioethics committee has been the NBAC (National Bioethics Advisory Commission), which played a key role in the debates over patenting, informed consent, and genetic engineering. In 2002, the NBAC's charter gave way to its replacement, the President's Council on Bioethics. On the top of its list—a list made by President George W. Bush—is human cloning, as well as stem cells and bioterrorism. For more, go to http://www.bioethics.gov.

2. Bruno Latour, *The Pasteurization of France* (Cambridge: Harvard University Press, 1988).

3. Latour, in another book, gives a simpler example—the speed bump. The speed bump is just a lump of concrete, perhaps painted. But in the context of local traffic, it signifies—physically signifies—"slow down." It fulfills the role a person such as a traffic cop might play if the speed bump were not there. Further, it specifies a relationship between particular people and things—between people driving cars, as opposed to pedestrians—and a local site where there are other cars and people. See Bruno Latour, *Pandora's Hope: Essays on the Reality of Science Studies* (Cambridge: Harvard University Press, 1999), 185–90.

4. Examples of this strand include Peter Singer, *Writings on the Ethical Life* (New York: Ecco, 2001); Michael Ruse, ed., *Biology and the Foundations of Ethics* (Cambridge: Cambridge University Press, 1999). To be fair, Singer's writings span the range of approaches to ethical and bioethical issues. His *Rethinking Life and Death* (New York: St. Martin's Press, 1996) attempts to make the claim, through a series of informal, close studies, that new medical technologies are requiring us to rethink our common notions of life, death, and nature.

5. Examples include Arthur Caplan and Daniel Coelho, eds., *The Ethics of Organ Transplants* (New York: Prometheus, 1999); Leon Kass and James Wilson, *The Ethics of Human Cloning* (New York: AEI, 1998); and Robert Baker, Linda Emanuel, Arthur Caplan, and Stephen Latham, eds., *The American Medical Ethics Revolution* (Baltimore: Johns Hopkins University Press, 1999). Applied bioethics studies tend to focus on a particular issue (human cloning, New Reproductive Technologies [NRTs], patenting), while more philosophically oriented approaches tend to forgo case studies in favor of a broad humanistic framework. Caplan works at the Center for Bioethics (University of Pennsylvania), among the most prominent academic research organizations pursuing applied bioethics approaches.

6. A number of case studies as well as more theoretical essays are contained in Peter Singer and Helga Kuhse, eds., *Bioethics: An Anthology* (London: Blackwell, 1999).

7. Examples include Donna Haraway, *Modest.Witness@Second_Millennium* (New York: Routledge, 1997); Colin Jones and Roy Porter, eds., *Reassessing Foucault: Power, Medicine, and the Body* (New York: Routledge, 1998); Dorothy Nelkin and Susan Lindee, *The DNA Mystique: The Gene as a Cultural Icon* (New York: W. H. Freeman, 1995); and Margrit Shildrick, *Leaky Bodies and Boundaries: Feminism, Postmodernism, and (Bio)ethics* (New York: Routledge, 1997).

8. See Michel Foucault, "What Is Critique?" in *The Politics of Truth* (New York: Semiotext[e], 1997).

9. For a historical view of eugenics movements in the United States and Europe, see Daniel Kevles, *In the Name of Eugenics: Genetics and the Uses of Human Heredity* (Cambridge: Harvard University Press, 1995). The examples of the Tuskegee study (syphilis studies on economically disenfranchised black men in Alabama, in which treatment was withheld and the disease allowed to run its course) and the Nuremberg Code (ethical codes established by the international community following the Nazi trials) are both examples of the extremes to which the total absence of bioethics can

lead. In a similar vein, the more recent controversy surrounding the Neem tree (natural resource in India, which was patented by more than a dozen U.S. and European corporations, without the consent of the community) places this lack of ethics in an economic context tied to uneven relations between nations. For more on the Neem tree, see Vandana Shiva, *Biopiracy* (Toronto: between the lines, 1997).

10. For more on the historical development of medical technologies, see Bettyann Holtzmann Kevles, *Naked to the Bone: Medical Imaging in the Twentieth Century* (Reading, Mass.: Addison-Wesley, 1997).

11. See Dion Farquhar, *The Other Machine: Discourse and Reproductive Technologies* (New York: Routledge, 1996); and Valerie Hartouni, *Cultural Conceptions: On Reproductive Technologies and the Making of Life* (Minneapolis: University of Minnesota Press, 1997).

12. See Jeremy Rifkin, *The Biotech Century: Harnessing the Gene and Remaking the World* (New York: Tarcher/Putnam, 1998).

13. Recall that Kant's "critical project," especially the *Critique of Pure Reason*, is in part a response to the problems posed by the skepticism of British empiricism. Prior to Kant, Hume pushed the ability to know the world "out there" to its limit, suggesting that the very notions of the object world and causality may be merely the result of habitual modes of thinking about the world. This crisis in ontology is brought into the domain of epistemology by Kant, who shifts the debate from the world out there to the subject mapping the world out there (the categories of understanding). For Kant, we may never know the world in itself ("noumena"), but what we can know is our knowing of the world ("phenomena"). To do this, according to Kant, we need to assess the limits and capabilities of the faculty of reason. The challenge in the position Kant outlines becomes more evident when considering ethics. How do we verify, account for, and most of all understand an other subject "out there"? Attempting to find a position between the extreme poles of idealism and empiricism, Kant suggests that morality—"good" and "evil" in relation to an other "out there"—must, in order for it to be considered moral, function according to an "unconditional rational necessity." In other words, morality can only be taken as valid if it is considered as universal (applicable for everyone) and rational (a priori or prior to particular experience). "Moral laws" are therefore a form (and not specific content) of what Kant terms the "categorical imperative" or the "ought" of ethical action.

14. Immanuel Kant, *Ethical Philosophy: The Complete Texts of Grounding for the Metaphysics of Morals and Metaphysical Principles of Virtue*, trans. James W. Ellington (Indianapolis: Hackett, 1983).

15. Ibid., 24, 26.

16. Ibid., 30.

17. This leads to a consideration of the ends involved in the unconditional rational necessity as a formal law. For Kant, this resides in defining the essential characteristics of being human, which for him is the power of rational self-determination. This is the Formula of Humanity: act always so that you treat humanity, in your own person or another, never merely as a means but also at the same time as an end in itself.

18. By "freedom" Kant does not mean total fulfillment of self-interest. Freedom has a negative and a positive definition. Freedom is negatively defined as the liberation from the enslavement to animal instinct, appetite, and unreflective desires. To be free, for Kant, is to realize one's consonance with the faculty of reason. Reason is the vehicle of realizing freedom because it moves the subject beyond mere being into a domain of awareness, into the gaining of knowledge through understanding.

19. Michel Foucault, "The Birth of Biopolitics," in *Ethics: Subjectivity and Truth*, ed. Paul Rabinow (New York: New Press, 1994).

20. There is no body that precedes ethics, no body that separately acts ethically or unethically. Conversely, there is no ethics without bodies that act or do not act ethically. Is constructionism bidirectional? Bodies construct ethics because of Spinoza's question of what a body can do. Ethics constructs bodies because of the biopolitical question of how to govern. Biomedicine is the middle point between them, suggesting answers to what a body can do, as well as to what counts as normative, healthy, diseased, aberrant.

21. Benedict de Spinoza, *The Ethics*, trans. R. H. M. Elwes (New York: Dover, 1955).

22. Gilles Deleuze, *Spinoza: Practical Philosophy*, trans. Robert Hurley (San Francisco: City Lights, 1988). The term "Deleuze's Spinoza" may seem excessively academic, but it serves to emphasize that the approach here is Deleuze's particular way of working through Spinoza. Some of Deleuze's emphases occupy little space in the *Ethics* (such as the comments on animals), while elements that are central to the *Ethics* Deleuze finds less interesting (such as the question of God's existence). In general, Deleuze's reading of Spinoza is valuable because it translates what could otherwise be seen as exclusively theological concerns into the domains of political ontology and even contemporary science (e.g., complexity's notion of emergence and Spinozist attributes).

23. Spinoza, like his contemporary Leibniz, begins with a first principle: one substance for all attributes. Spinoza's metaphysics views the world as a set of innumerable instances of a single prime matter, which Spinoza variously calls "substance," "nature," and "God." (It is the latter association of theology with matter that gets Spinoza in trouble with the religious politics of Jewish Amsterdam in the 1650s.) In Spinoza's cosmology, there is first substance (prime matter) that is manifested in "attributes" (two of which we can intuit—thought and extension), which have at any given time "modes" (or states of transformation or nontransformation). For example, substance in extension such as wood (we can never know "substance" in itself, because by definition it in-forms the physical and mental worlds) can be formed into a chair, a book, or a pencil. Substance in extension in form, such as a chair, can have the attributes of being sturdy or broken. If the chair is in the process of being upholstered, its mode is specific to the attributes being transformed, which may have to do with more than fabric or color (from natural wood to black leather, from a desk chair to an S/M prop). The quality of the modes (possible transformation of a given state of an attribute) is what Spinoza calls "affects," which include, but are not limited to, human emotion. It is in this particular mode of affects that Spinoza, and Deleuze, find the roots of an ethical understanding that works with human modalities and relations, rather than autonomous human subjects.

24. Spinoza, *The Ethics*, 93 (II, "Digression on the Nature of Bodies," Ax.I, Lem.I).

25. Spinoza's doctrine of "parallelism" suggests that Cartesian dualism is mistaken; mind and body are not dualist but parallel. One substance, many attributes, two of which are thought and extension. It is the qualities of these two extensions that Cartesianism has misinterpreted: "I can scarcely believe, until the fact is proved by experience, that men can be induced to consider the question calmly and fairly, so firmly are they convinced that it is merely at the bidding of the mind, that the body is set in motion or at rest, or performs a variety of actions depending solely on the mind's will or the exercise of thought. However, no one has hitherto laid down the limits to the powers of the body, that is, no one has as yet been taught by experience what the body can accomplish solely by the laws of nature, in so far as she is regarded as extension. No one hitherto has gained such an accurate knowledge of the bodily mechanism, that he can explain all its functions; nor need I call attention to the fact that many actions are observed in the lower animals, which far transcend human sagacity, and that somnambulists do many things in their sleep, which they would not venture to do when awake: these instances are enough to show, that the body can by the sole laws of its nature do many things which the mind wonders at" (Spinoza, *The Ethics*, 132).

26. "In this way, Ethics, which is to say, a typology of immanent modes of existence, replaces Morality, which always refers existence to transcendent values" (Deleuze, *Spinoza*, 23).

27. Deleuze's example is metabolism, where bodies of food particles enter into relations of composition with bodies of cells in the digestive system, which are in turn connected to the anatomical body. See Deleuze's *Spinoza*, 22–26, 32–33.

28. Ibid., 22–25.

29. "Thus, animals are defined less by the abstract notions of genus and species than by a capacity for being affected, by the affections of which they are 'capable,' by the excitations to which they react within the limits of their capability. Consideration of genera and species still implies a 'morality,' whereas the Ethics is an ethology which, with regard to men and animals, in each case only considers their capacity for being affected" (ibid., 27).

30. This is implicit in Aristotle's notion of the greatest good for the greatest number, and it is formulated concisely in Kant's categorical imperative.

31. Humberto Maturana, "Metadesign," in *Technomorphica*, ed. Joke Brouwer and Carla Hoek-endijk (Rotterdam: V2, 1997), 167–203.

32. Although Maturana does not go into depth in his use of the terms *historical* and *ahistorical*, there is a sense in which the terms are not limited to the biological domain. Historical seems to mean different types of biological change (within and across species; individual versus group evolution), but it seems to apply to human cultures as well. Ahistorical change, not having a correlative of biological history (that is indeed its definition), is limited to its being a product of human activity—an extension of human bodyhood. But if this is the case, it would seem that technology/robots are indeed historical by virtue of their being tied to the domain of human activity. It may be that Maturana draws too severe a boundary between the human and robot, living systems and technical systems. Jean Baudrillard's characterization of the robot and the android blurs this boundary, just as the discourse of the "cyborg" does. See Jean Baudrillard, *Simulations* (New York: Semiotext[e], 1983); Chris Hables Gray, ed., *The Cyborg Handbook* (New York: Routledge, 1995).

33. This is another way of talking about "biomedia," the point at which the body is reconfigured in technical terms. However, in a way, biomedia is the inverse of Maturana's scenario. For Maturana, the recursive property of technology folding back on the human is technology reconfiguring the human, replaying the familiar anxieties associated with technological dehumanization. In biomedia, by contrast, technology does not "do" anything to the human; rather, technology is seen to inhere in the biological itself. Nothing is "applied" from one domain to another; instead, a particular ontology—a technical worldview—is embedded in novel systems, resulting in machine versions of "wet" flesh (prosthetics), or in new ways of thinking about technology altogether (DNA chips).

34. Maturana, "Metadesign," 187.

35. Ibid., 175.

36. The other of ethics is modes of praxis that simultaneously challenge the assumptions in currently existing ethical models; it is therefore critical. The other of ethics also forms relations between different ethical paradigms, including the relation between theory and practice; it is therefore pragmatic. Finally, the other of ethics takes philosophical-ethical investigations as a means toward investigating how our notions of "body," "self," and biological "life" are being transformed in light of developments in technoscience; it is therefore creative.

Index

Adleman, Leonard, 91

Affect, 186, 188

Anatomy, 160

Artificial intelligence (AI), 33, 101, 106–7; biocomputing and, 106–7; brain and, 103

Artificial life (a-life), 66

ASCII (American Standard Code for Information Interchange), 42

Autocatalysis, 164, 166. *See also* Boolean genetic networks

Autopoiesis, 168–72; allopoietic vs. autopoietic systems in, 192. *See also* System/systems

Baldi, Pierre, 41

Base pair complementarity, 15, 17, 53, 77, 79, 80–81, 94, 108, 113, 130, 198n34

Beadle, George, 37

Beadle, Muriel, 37

Benjamin, Walter, 7

Bergson, Henri, 60–61, 62, 205n56, 211n42

Bertalanffy, Ludwig von, 143–44. *See also* Systems theory

Biochips. *See* DNA chips

Biocomputing, 3, 30, 66; Adleman's proof-of-concept experiment, 94–97; artificial intelligence and, 106–7; bioinformatics and, 3–6, 89, 98; computability and, 99–101; definition of, 89; description of, 90–91; networks in, 107–9, 112–13; time and, 109–11

Bioethics, 177–80, 188; categorical imperative (Kant) and, 182–84, 186; compared to "bio-ethics," 184–91; cultural aspects, 175–76; death

and, 179; ethics without morality (Deleuze) and, 186, 190; informed consent and, 178–79, Kantian moral law and, 181, 183–84, 189

Bioinformatics, 2–3, 19, 21, 30, 201n6; biocomputing and, 3–6, 89, 98; as a business, 35; companies in, 35–36; databases in, 35, 187–88; definition of, 33, 35; description of, 34–37; IT sector interests in, 35; materiality and, 41–45; open source and, 56–60; scripts and, 55; sequence-structure division and, 158; techniques in, 19–22, 36–37, 46–50. *See also* Biological data; Biomedia

Bio-logic, 22, 44, 47, 62, 85

Biological computing. *See* Biocomputing

Biological data, 42–46, 53–54, 159. *See also* Genetic code

Biological information. *See* Biological data

Biological phenomenology, 31, 172–73

Biology. *See* Biomedia; Genetic code; Molecular biology

Biomedia, 1–31; biocomputing and, 96–97; body-technology relationship and, 13–15; consequences of, 26–28; decoding in 22–26; definition of, 5–7, 11, 28, 57, 63, 81, 157; DNA chips and, 80; encoding in 16–18; recoding in 18–22; total translatability in, 50–51. *See also* Body/bodies

BioMEMS, 65–67; application and types of, 67–71

Biomolecular body, 41, 51–52, 161–63

BioPerl, 58–60

Biopolitics, 183

Biosensors. *See* BioMEMS

Eugene Thacker is assistant professor of new media in the School of Literature, Communication, and Culture at the Georgia Institute of Technology. His writing on the social and cultural aspects of biotechnology has been published widely and translated into a dozen languages.

Printed in the USA
CPSIA information can be obtained
at www.ICGtesting.com
JSHW060515250124
55732JS00009B/44